STABILITY AND CHANGE

Innovation in an Educational Context

ENVIRONMENT, DEVELOPMENT, AND PUBLIC POLICY

A series of volumes under the general editorship of
Lawrence Susskind, *Massachusetts Institute of Technology, Cambridge, Massachusetts*

PUBLIC POLICY AND SOCIAL SERVICES
Series Editor:
Gary Marx, *Massachusetts Institute of Technology, Cambridge, Massachusetts*

STABILITY AND CHANGE: Innovation in an Educational Context
Sheila Rosenblum and Karen Seashore Louis

FATAL REMEDIES: The Ironies of Social Intervention
Sam D. Sieber

Other subseries:

ENVIRONMENTAL POLICY AND PLANNING
Series Editor:
Lawrence Susskind, *Massachusetts Institute of Technology, Cambridge, Massachusetts*

CITIES AND DEVELOPMENT
Series Editor:
Lloyd Rodwin, *Massachusetts Institute of Technology, Cambridge, Massachusetts*

STABILITY AND CHANGE

Innovation in an Educational Context

Sheila Rosenblum
and
Karen Seashore Louis

Abt Associates
Cambridge, Massachusetts

With the assistance of
Nancy Brigham and Robert E. Herriott

Plenum Press • New York and London

Library of Congress Cataloging in Publication Data

Rosenblum, Sheila.
Stability and change.

(Environment, development, and public policy: Public policy and social services)
Bibliography: p.
Includes index.
1. Education, Rural—United States—Planning. I. Louis, Karen Seashore, joint author. II. Title. III. Series.
LC5146.R67 370.19'346'0973 80-28291
ISBN 0-306-40665-9

A Division of Plenum Publishing Corporation
233 Spring Street, New York, N.Y. 10013

Printed in the United States of America

Foreword

Nearly a century ago, Emile Durkheim founded the sociology of education on the French cultural and structural premise that the function of educators is to transmit culture from one generation to the next. The clarity of his vision was aided by the era, the place, and the actors in the learning environment. His was an era when the relatively seamless web of western culture, although ripping and straining, was still intact. The place, post-Napoleonic France, was vertically stratified and elaborately structured. And the teachers had reason to think they were agents of authority, whereas most students, during school hours at least, behaved as if they were the objects of that authority. Underlying the very notion of a sociology of education, then, was a visible and pervasive aura of a system and order that was culturally prescribed.

Scholars of American education have yearned for such systems before and since Durkheim. Every European and English model has been emulated in a more or less winsome manner, from the Boston Latin School of the 1700s to the Open Education programs of the 1960s. In the last quarter century of research, it has begun to dawn on us, however, that no matter how hard American educators try, they do not build a system. They occupy nonsystems because western, and particularly North American, culture is too fragmented and diffuse to be transmitted reliably or uniformly; because education was deliberately omitted from the United States Constitution; because stratification designs shift and dissolve in the only universal societal solvent, which is money; and because American individualism, privatism, and localism are not the soil from which pedagogical authority and its urgently essential reciprocal of obedience appear to spring with much vitality.

When James Coleman and his associates reported in 1966 on the state of public education in the United States, the nonsystem properties of public schools were so abundant that their landmark study attributed about 3% of all student variations in school learning to teaching influences, compared to more than 60% to the factors of social class and ethnic origin. If public schools seemed to have systems properties, they reasoned, they worked rather like August Hollingshead and William Lloyd Warner had found them working thirty years earlier: as sorting or sifting devices and as peer mixers that socialized youth toward a vestibular version of adult community.

Nonetheless, Coleman and his associates also reported that most of the schools in their vast sample seemed to them to be strikingly standardized and interchangeable as to curriculum, facilities, staff, and operating procedures, irrespective of great external regional, state, and local differences. Their four hundred sample districts displayed school patterns so common as to facilitate great ease of student transfer from one locality to another. Seymour Sarason illuminated this pattern later by explaining the sense in which teachers have accrued and maintain a definite *school culture* that can be distinguished from other parts of the culture at large, and he viewed students as transient sharers or samplers of this culture.

It has fallen to Sheila Rosenblum and Karen Seashore Louis, however, to develop a set of concepts and a corresponding body of evidence that is patterned credibly by those concepts for studying a series of educational entitites *as if* they were the elements of a system. Theirs is not an exercise in writing fables, nor have they imposed system and order on a nonsystem for the delivery of educational services. Instead, their concepts are grounded in social-system theory, yet they are designed to fit and to help us account for empirically authentic regularities in the behavior of administrators, teachers, school boards, and state and federal agents who make policies and fund programs in concert with teachers.

Their concepts are suitably fluid. They come out of several disciplines and thus disclose more properties than can be found by using the narrow lense of sociology. Their concepts, when applied to the entities and events of ten Experimental School Projects funded by the National

Institute of Education, reconcile otherwise scattered, divergent, and seemingly random actions. Through their very wide angle lenses, we see how an apparently unconnected and disparate set of entities—schools, local districts, state agencies—can be interpreted meaningfully as interacting parts of an extraordinarily "loosely linked" yet reciprocating system.

The Rosenblum–Louis concepts are not original. They are derived from carefully developed ideas about organizational behavior, cultures, and change processes as these have been researched for decades in social psychology, anthropology, sociology, and the administrative and political sciences. What is both original and invaluable about this book is the extent to which concepts have been selected, refined, and fit together in ways that fulfill the basic purposes of scientific research: they increase our ability to predict, control intellectually, and thus even to plan future social actions!

Their concepts of system linkage, refined into a logical set of types amenable to measurement, equip us to make sense out of crucial features of institutional formation and transformation under North American conditions. With their typology and their approaches to measuring interpersonal relations within schools and structural relations in the external milieu, we have the means at hand for *unifying* some essential approaches to educational research. Previously we were stranded in the macrosocial approaches of most sociologists or confined to excessively "intensive" studies of social process within the isolative walls of separate schools. In my estimation, their book advances the explored frontiers where behavioral and social organizational researchers often fail to band together and expand our ability to know.

Rosenblum and Louis map a part of that frontier, but their imaginative expedition is—in keeping with the fate of most explorers of American educational frontiers—hampered by the *absence* of sound advance preparation by those who designed the Experimental Schools program at the onset of the decade of the 1970s. Their approach is capable of extending organizational inquiry to include the measurement of learning outcomes for students, I believe, but the program they researched was itself too poorly designed to include definite pedagogical interventions. In a companion monograph devoted to the measurement and analysis of student effects in the Experimental Schools, Wendy Peter Abt and Jay Magidson (1980) applied impressive methods but came up, inevitably it would seem, with few positive findings.

Weaknesses in program and instructional designs paradoxically become an event for analysis by Rosenblum and Louis, rather than a source of negative findings. In this sense, their "theory of the middle range" has impressive flexibility, and I look foward to future inquiries in which both stability and change in schools can be connected more tellingly to the life chances or well-being of students.

The theoretical perspective of Rosenblum and Louis thus takes us deeper into the culture of schooling than did the work of the macrosociologists. It helps explain the seemingly opaque organizational features of the American public school nonsystem. The patterned regularities the authors identify grace the teaching and learning environment with reason. More than superficial characteristics are shared, and when these are made more widely understood, they can assist us in planning changes in the learning process itself.

Theirs is a lucid, coherent, and practically valuable contribution to the study of change. I expect this book to become an enduring chart for many future explorers of the cross-disciplinary frontier of organizational analysis.

ROBERT A. DENTLER

Preface

This volume examines the organizational characteristics of school systems that affect the process and outcomes of planned change activities. It is written for organizational and educational researchers, and those persons interested in the design and management of educational change programs at the federal, state, and local levels. Our analysis is the product of a multiyear study of a set of educational field experiments known as the Rural Experimental Schools (ES) program, which was funded by the National Institute of Education, and which involved 10 geographically dispersed school districts comprising 52 schools. These school systems were, according to guidelines articulated by the federal programs designers, to develop and implement plans for "comprehensive change," which was to affect all aspects of district and school functioning.

Although the Rural ES program never elaborately defined "comprehensiveness," any project funded by ES had to involve all schools, all pupils, and all subject areas in a district, and had to involve five "facts of comprehensiveness": (1) a fresh approach to the nature and substance of the total curriculum in the light of local needs and goals; (2) reorganization and training of staff to meet particular project goals; (3) innovative use of time, space, and facilities; (4) active community involvement in developing, operating, and evaluating the proposed project; and (5) an administrative and organizational structure to support the project.

The notion of planned comprehensive change is not characteristic of the way in which innovation has typically been approached in schools or school districts. In general, change in schools may be best characterized as not only "piecemeal" (that is, school problems are dealt with sequentially as they become particularly pressing or obvious), but localized within individual classrooms or schools. The notion of comprehensive change, on the other hand, called for districtwide and "holistic" change of all parts simultaneously in an integrated fashion throughout all schools. Comprehensiveness was a federally mandated prescription for change rather than a goal that arose from felt needs on the part of the 10 participating school districts. The diffuseness and vagueness of the concept of "comprehensiveness" created obstacles to success, and implied an approach to change which was, in most cases, at odds with local agendas. ES as a program had a two-pronged approach. One was comprehensive change; the other was a locally planned change program that was developed in response to locally perceived needs and problems. The duality of locally planned change on the one hand, and a loosely defined federal mandate for comprehensiveness on the other, came into inevitable conflict. In the end, all parties felt, to some degree, "cheated"—the local administrators and teachers because they thought they were involved in a program that would meet their own needs and problems "with no strings attached," and the federal sponsors because they thought the local school districts had agreed to plan and implement "comprehensive change." As the program filtered through many layers and levels—from the federal office to the school buildings to the classroom teachers—multiple redefinitions of the situation often caused misunderstanding and distortion.

Despite such problems, many changes were implemented and are being sustained in the ES districts. The attempt to change many or all parts of the organizations' functioning brought about an accumulation of small successes. Several important factors, such as the characteristics of the authority structure and measures of organizational climate, acted as facilitors or as barriers to the process of change.

Our work examines in great detail the relationship between organizational properties of schools and school districts, and the outcomes of innovative programs. Particular attention is paid not only to the organizational properties of parts of the educational system, such as individual schools, school districts, and federal educational structures, but to relationships between system parts, or what we term "system linkages." Our findings suggest that the nature and strength of the ties between various parts of the system are critical elements in explaining the implementation of new programs.

The research upon which this volume is based was part of a larger, seven-year effort conducted by Abt Associates Inc. under contract to the National Institute of Education. The research approach of the project consisted of a study of the rural school districts through a combination of full-time participant observation within each district, periodic testing of students, and a series of surveys and interviews with students, teachers, administrators, and community residents. A long-term research effort owes a debt of gratitude to many people. We particularly wish to acknowledge the many members of the Abt Associates staff who contributed to the preparation of this volume: first and foremost the project director, Robert E. Herriott, who initiated and guided this part of the larger research effort from its inception. Nancy Brigham was the major author of Chapters 4 and 5 and spent many hours mining the site case studies for use in interpreting and enlivening our analyses. We also acknowledge the contributions of James Molitor, JoAnn Jastrzab, John Clark, James Hoyle, and Lisa Lang. The support of our colleagues Wendy Abt, Michael Kane, Don Muse, Steven Fitzsimmons, Richard Anderson, and James Vanecko was also greatly appreciated. Our gratitude is extended to Russell Mullens, Lynne Fender, Robert Cunningham, Susan Abramowitz, and Norman Gold, who served as project officers and reviewers for the National Institute of Education and whose advice, support, and criticism greatly facilitated the conduct of the study.

This volume was greatly enriched by the efforts of 11 "on-site researchers," whose primary research materials and whose case studies provide special insight into the uniqueness of each of the school districts under study. They are: Allan F. Burns, Charles A. Clinton, Carol Pierce Colfer, A. Michael Colfer, William L. Donnelly, William A. Firestone, Lawrence Hennigh, Stephen J. Langdon, Donald A. Messerschmidt, Charles I. Stannard, and C. Thompson Wacaster. We are also grateful for the contribution of consultants who provided insights and guidance at various stages in this research process: Professors W. W. Charters, Jr., of the University of Oregon, Ronald Corwin of Ohio State University, Terrence Deal of Harvard University, Alan K. Gaynor of Boston University, Neal Gross of the University of Pennsylvania, Benjamin Hodgkins of the University of

Manitoba, and Charles Perrow of the State University of New York, Stony Brook.

This volume could not have been completed without the superb office skills of Timothy Burns, Lillian Christmas, and Karen Malmberg. We especially acknowledge the contribution of Peter Desmond and Mary Schumacher, whose editorial assistance greatly benefited this manuscipt.

We are also indebted to the teachers, principals, superintendents, project directors, and other administrators in the 10 school districts who agreed to have us "look over their shoulders" and question them, often at length. They have assisted us generously, despite the valuable time it took away from their primary task of improving their educational programs. We are grateful for their participation.

This work was supported by the National Institute of Education, Department of Education, and we are particularly appreciative of the support of the Research and Educational Practice Program under whose sponsorship the final stages of the work was conducted. However, the content does not necessarily reflect the position or policy of that agency, and no formal endorsement should be inferred.

In the case of joint authorship, it is sometimes easy to define the precise contribution of each partner. In this case it is impossible. This has truly been a collaborative effort, through which friendship and mutual respect have grown. We jointly thank all the above—and many others too numerous to mention—who contributed greatly to the success of the research.

SHEILA ROSENBLUM
KAREN SEASHORE LOUIS

Contents

5 THE BEGINNING OF CHANGE: THE PLANNING YEAR 95

6 THE IMPLEMENTATION OF PLANNED CHANGE IN SCHOOLS 127

7 FURTHER EXPLORATION OF IMPLEMENTATION IN SCHOOLS 153

1

Introduction

All change involves risks, but for the contemporary American school, the "safe" strategy of maintaining old structures and yesterday's curricula is often a poor choice. Declining enrollments, rapid changes in the existing technology and knowledge about teaching and learning processes, a continual expansion of the role of the school into new areas, and changes in the prevailing cultural preferences of both local communities and the larger society continually impel schools to innovate.

Yet, although these pressures create a climate which favors the introduction of planned programs of change, many, if not most, of the efforts of schools at planned change fail to achieve either their stated or implicit goals (see, e.g., Charters & Pellegrin, 1972; Gross, Giacquinta, & Bernstein, 1971; Herriott & Gross, 1979). One prominent explanation for the persistent finding of failure is the poor understanding among school personnel and policy makers of the organizational properties and processes that may have a significant impact upon a school's ability to achieve the designed outcomes of such change programs.

Pressures for change have not affected all schools to the same extent. With greater or lesser success, urban and suburban schools have, in many cases, become experienced in grappling with new programs such as individualized instruction, modular scheduling, computer-assisted instruction, differentiated staffing, and experience-based career education, to name only a few of the currently favored educational innovations. In the rural schools of America, however, the pace and pattern of change have been quite different. Although rural areas have had to confront many changes associated with school-district consolidation, they have been less affected by much of the environmental turbulence and cultural disorientation of the sixties. Rural areas have only recently encountered many of the pressures for change that are now taken for granted in most urban and suburban school systems in the United States.

In July 1971, the U.S. Office of Education began an innovative program of research and development to address the issue of school improvement. The purpose of the Experimental Schools (ES) program, as it was called, was to understand better how communities can be helped to mobilize their resources in order to upgrade education. The ES program was also intended to help bridge the gap between educational research and practice by directing the attention of researchers to problems which local officials consider to be most critical. The program, which was transferred to the new National Institute of Education (NIE) in 1972, sought to test the hypothesis that

> significant and lasting improvements in education, beyond those made possible by piecemeal innovative elements, are more likely to occur if comprehensive changes are introduced into all elements of a school system. (National Instutute of Education, 1974, p. 38)

Although initially made available only to large-city school systems, in 1972 the program was extended to rural school districts. What follows is an account of how 10 diverse rural school districts selected for participation in this federal program responded to this opportunity to improve their schools. In addition, we hope to extend the social-science literature concerned with the process of planned change by drawing upon the varied experiences of the 10 districts as they charted their courses through a five-year process of planned change.

WHY STUDY CHANGE IN RURAL SCHOOLS?

Rural schools merit attention because they have more than their share of problems. Although social-policy planning has become an accepted part of American society, programs relevant to *rural* educational needs and to comprehensive solutions of rural problems received very limited attention until the late 1960s. Federal outlays for rural areas in 1973 still constituted only 11% of library and materials funds, 13% of basic vocational aid, 8% of migrant education aid, and 20% of bilingual education funds (U.S. Senate, 1975). Needs in rural areas are many; over 14 million persons in such areas fall below the poverty level. The depressed economies characteristic of such areas have discouraged investments by the private sector, thus reinforcing chronic employment problems. Local rural governments generally lack the industrial tax base to finance many social services, as well as the skilled manpower to plan and coordinate programs of social progress. In rural areas there is also a pressing need for better-trained teachers, more effective administrators, longer school terms, more adequate school plants, better use of consolidated school plants as social centers, better programs for adults, more attention to the health problems of children, and improved educational programs based on the social needs of rural children and dedicated to the improvement of their mental and physical growth (Smith & Zopf, 1970).

However, there are other reasons for studying educational change in rural sectors. The study of rural schools can help us understand organizational and school behavior in general. The fact that rural schools have lagged behind other schools in making major changes in their structures and curricula makes them a very appropriate locus for studying the processes and outcomes associated with innovative programs. In contrast, the study of innovation and change in urban and many sububan schools is made difficult by the multiplicity of established or recent innovative programs. In schools that frequently have several active innovative programs annually, it is often impossible to isolate any one program from others, either financially or in terms of actual activities. In addition, the rapid sequencing and overlap of innovations often desensitizes staff members to the process of innovation. It may even affect the research that frequently accompanies new programs. An outstanding example of this occurred recently when a sociologist, who was conducting a case study of the process by which a school in an urban area was adopting a new program, requested an interview with a teacher who was reported to be active in the implementation process. "Oh," replied the teacher, "are you the sociologist studying X or Y program? I've been asked to be interviewed about both of them this week."

Another reason for studying change in rural America emerges from the nature of the educational systems found there. Rural school districts tend to be smaller and somewhat less organizationally complex than urban schools. As such, they permit the researcher to form a more complete picture of their organizational properties and of their interactions, both internal and with external forces, than is possible in many urban settings.

This argument is, of course, only as compelling as the evidence that suggests that rural schools and their communities are different primarily in degree rather than in kind from other school systems. While not overlooking the fact that some issues and problems are more salient in rural areas, we would argue that the similarities between rural and other schools are, in many ways, far greater than the differences. Structurally, for example, rural schools very much resemble other American schools. They have school boards, superintendents, teachers, and a variety of specialists to serve the schools. There is a district office and a number of schools arranged in levels—often elementary, junior high, and high schools, but occasionally exhibiting different patterns. They face many problems of funding and of acquiring and utilizing resources, problems that are familiar to all but the most affluent school districts. Unlike the popular image, they are not necessarily homogeneous organizations with high morale and strong relationships with their communities. Rather, like schools in more densely populated areas, they are often plagued by controversy, strikes, and complaints about the degree to which they reflect community values and norms.

In sum, although rural schools have their own unique problems that deserve attention, as organizations they are in most ways quite similar to other American schools. A study of the 10 districts which took part in the ES program can thus provide insights into the problems of educational change that extend far beyond the rural milieu.

THE FEDERAL ROLE IN EDUCATION

The ES program represented a major response by the federal government to the needs of rural areas and a major attempt to address pressing inequities in education. It marked a distinct point on the continuum of federal involvement with education at the local level. Prior to the 1950s, the federal government's role in education was largely passive: the U.S. Office of Education's primary function was to collect statistics on schools and schooling. In 1950, the federal government bore only 2.8% of the costs of elementary and secondary schooling; by 1975, in contrast, this figure had risen to 9%. This increase represents only part of the picture, however, for the change in the federal role has been as much in influence as in dollars.

In earlier years, federal aid to education was in the form of limited funding distributed to the local school districts as general aid or "block grants." However, in the Sputnik period, when people became concerned with the comparative excellence of American schooling, the federal government assumed a more important role in improving the content and structure of the educational system (Bailey & Mosher, 1968; Kirst, 1973; Sieber, 1974). Rather than providing limited funds to subsidize existing costs of education, the government gave the local districts more aid; but with the added dollars came increased federal influence in the schools. The shift from general aid to "categorical aid" was promoted through federal legislation such as the National Defense Education Act (NDEA) in 1958, and the Elementary and Secondary Education Act (ESEA) in 1965.

Because the American educational system is composed, at least in theory, of locally autonomous school districts, the federal government did not simply mandate changes. Rather, federal agencies tried to become partners in change by funding research projects and demonstration programs designed to produce improved methods and curricula, which could then be diffused to the wider population. Increasingly, the federal government has chosen to enter into contractual relationships with local education agencies or various intermediate education agencies to carry out change programs that are actually designed by Washington policy makers.

By the end of the 1960s, the results of large and costly efforts in educational change instituted by both the federal government and private agencies had proven disappointing to their sponsors. The Ford Foundation studied its own $34 million investment in educational change and found "little if any residual effect from any of the projects only three years after they were completed" (Ford Foundation, 1972). In addition, a recently completed study of categorical grant programs sponsored by the Office of Education has concluded that few schools show any permanent effects from the massive expenditures involved in these "seed-money" efforts (Berman & McLaughlin, 1977).

Neither general nor categorical aid seemed to have paid off. But the various programs had something in common: they had usually attempted to change aspects of the school in a piecemeal fashion, and they had usually failed to improve educational outcomes (Baldridge & Deal, 1975).

At the same time, evidence emerged that traditional research-and-development models generated in industrial settings might not be appropriate for educational innovation and change (Kirst, 1973; Louis & Sieber, 1979). No longer was it assumed that a standard, centrally developed product could be applied to all local-school contexts. Rather, there was a growing emphasis on the need to adapt innovations to perceived problems and needs at the *local* level, and to allow adaptations of change programs to unique local environments. Furthermore, while the changes in the federal role have been noted by a variety of observers, and their impacts upon state departments of education have been well documented (Milstein, 1977), little attention has been paid to the role of the federal government as a direct partner in the change process, as opposed to a source of funding for the change process at the local level.

BASIC PREMISES OF THE RURAL EXPERIMENTAL SCHOOLS PROGRAM

Those seeking to explain the failure of federal programs to achieve desired impacts concluded that

> There are interrelationships between and among *all* the components of schooling. Furthermore, formal education actually amounts to only a limited intervention in the lives of individuals. Thus, change which affects only one educational component or one part of a school day represents so slight an impact on a child's total experience that dramatic effects are unlikely.
>
> Limited knowledge exists about how to improve the outcomes of a formal educational structure; efforts towards educational change may not be focusing upon what are actually the significant components of the educational process. (Budding, 1972)

A new approach to planned educational change was needed which would take into account these conclusions. What resulted was the four-pronged philosophy of the ES program:

1. *Local planning.* Locally stimulated program changes matching perceived needs were thought to be more likely to be continued through local funding than externally designed programs.

2. *Comprehensive change*. Educational programs which address all parts of the educational system simultaneously were thought to be more effective than past efforts which focused on piecemeal change.[1]
3. *Five-year terminal funding*. Funding was promised for a period of five years to allow changes to be adequately planned, implemented, and routinized. No fixed amount was projected, but awards were to be based on unique needs. However, no single grant would exceed 15% of the annual operation costs of the system(s) involved.
4. *Comprehensive research*. The ES program was intended to provide the opportunity for research in the process of planned educational change and in the significant components of the educational process.

The ES program assumed a dynamic partnership between the federal government and local school systems on the one hand, and the federal government and a research team on the other, providing a timely opportunity to study the interrelation between program sponsors and the enactors to change. The research reported in this volume is one product of that effort.[2]

THE RESEARCH CONTEXT

One of the basic premises on which the Rural ES program was established was that further research on educational change and edu-

[1]Although the Rural ES program never put into writing an elaborate definition of the concept of comprehensiveness, its staff members generally led school districts to understand that, in order to be comprehensive, a project must involve all schools, all pupils, and all subject areas; in particular, it must involve five concerns, which became known as "facets" of comprehensiveness: (1) *curriculum*—a fresh approach to the nature and substance of the total curriculum in the light of local needs and goals; (2) *staffing*—reorganization and training staff to meet particular project goals; (3) *time, space, and facilities*—innovative use of time, space, and facilities; (4) *community participation*—active community involvement in developing, operating, and evaluating the proposed project; and (5) *administration and organization*—an administrative and organizational structure to support the project.

[2]Other products of this research effort include Abt and Magidson (1980), A. Burns (1979), Clinton (1979), Colfer and Colfer (1979), Corwin (1977), Donnelly (1979), Firestone (1980), Fitzsimmons and Freedman (1981), Fitzsimmons, Wolff, & Freedman (1975), Hennigh (1979), Herriott and Gross (1979), Messerschmidt (1979), Stannard (1979), and interim reports that have preceded this volume: Herriott and Rosenblum (1976), and Rosenblum and Louis (1977).

cational processes was needed. Little knowedge existed concerning either the kinds of educational intervention that would produce desired effects or the process of planned change itself. If planned educational interventions were or were not working, why was this so? What are the organizational and envionmental constraints or facilitators of change?

The knowledge base was sparse, in part because few large-scale change efforts had been undertaken in the past, and in part because the situation reflected the "state of the art" of the study of the educational change process. These limitations were vexing, both to the designers of the Rural ES program (cf. Budding, 1972) and to us, particularly as we moved toward exploring and analyzing the implementation of the planned changes. In the remainder of this section, we point out weaknesses in the knowledge base and discuss recent trends in research to describe the context for our study.

Limitations of the Literature on Educational Change

Three limitations characterized research on planned educational change in the early 1970s. First, the research preoccupied with adoption of change by *individuals* rather than by organizations. Second, even the organizational studies were more concerned with *adoption and planning* of change than with its implementation. Finally, preoccupation with the *fact* of implementation (and its definition) tended to obscure the importance of the degree to which implementation was actually achieved.

Preoccupation with Individuals

The study of change has roots both in the human relations school of social research and in rural sociology. Human relations theorists have generally focused on individual adopters of innovations and the importance of individual leadership in promoting adoption of change. This individualistic bias leads to an emphasis on personal attitudes and characteristics, thus equating the process of change with a change in individual attitudes. (For corroboration, see Baldridge & Deal, 1975.) The *organization* as adopted has tended to be ignored.

Rural sociologists have also approached the process of change with a focus on the individual. Their work, which has a product orientation, generally deals with the adoption of such technically simple innovations as hybrid seed corn by individual farmers in a particular region (Gross *et al.*, 1971). Thus, Rogers's (1962) review of 506 studies of adoption and diffusion draws no conclusions about the organization as the adopter of innovation or about organizational features affecting the process of change.

However, the research of the 1970s has had to deal with innovations which were to be adopted by organizations rather than by single individuals, and the social-science literature is beginning to reflect this fact. There is a trend away from the individual "buyer/adopter"; now organizations and their constituent parts are being examined as units undergoing change. Educational researchers have begun to explore the interrelationship of the process of change and the organizational context in which the changes take place (Deal, Meyer, & Scott, 1975; Greenwood, Mann, & McLaughlin, 1975). Many such studies deal with organizational constraints on individual propensities for change (Bentzen, 1974; Gross *et al.*, 1971). Greenwood *et al.* (1975), in particular, have called for researchers to look beyond the details of the innovative project under study and focus on characteristics of the complex organization in which innovation is being carried out.

Preoccupation with Adoption and Planning

Rural sociologists, with their emphasis on individuals, have also tended to see adoption as the end of the change process. Thus, Rogers (1962) defines five stages in the change process; the last of these is adoption. But even the growing number of organizational change studies have been preoccupied with adoption and planning of change and have tended to overlook important issues of implementation. (For corroboration, see Gaynor, 1975; Gross *et al.*, 1971.) For instance, many studies emphasize the importance of change agents and of participation of "subordinates" for successful initation (Argyris, 1972; Bennis, 1966; Havelock, Guskin, Frohman, Havelock, Hill, & Huber, 1971). Other researchers emphasize the importance of the making of decisions for planned change by administrators (Bishop, 1961; Heathers, 1967). However, what degree of participation by various groups in an organization (and at what stage in the change process) will optimally affect the change process remains unclear (French, Israel, & Dagfinn, 1960; Morse & Reimer, 1956).

The preoccupation with early stages of the change process is due in part to the lack of models suitable for the study of change in organizational settings (Charters & Jones, 1975). However, the unsuccessful or disappointing outcomes of the extensive social and educational programs funded by the federal government in the 1960s have brought the realization that planning, designing, and disseminating programs of change do not guarantee their being implemented (Greenwood *et al.*, 1975). For example, despite growing interest in and support for such innovations as ungraded classes and team teaching, these innovations have rarely been successfully implemented. School districts have tended to *adopt* ungraded classes and then essentially *implement* a graded curriculum in the ungraded class (Pincus, 1974).

Thus, although the problems of implentation plague the innovation process and limit its outcomes, confusion about the process of implementation abounds, with insufficient analytical understanding available in the educational or organizational literature (Mann, 1976).

Preoccupation with the Fact Rather than with the Degree of Implementation

For the most part, the existing literature on implementation has concerned itself with the *fact* of implementation, its nominal definition, its identification as a stage in the planning process, and with the facilitators and barriers to implementation (Gross *et al.*, 1971; Hage & Aiken, 1970; Zaltman, Duncan, & Holbek, 1973). Studies have also focused on failures to implement planned innovations (Charters & Pellegrin, 1972; L. Smith and Keith, 1971), or on small successes (Baldridge & Deal, 1975).

However, there is no way to judge the effects of an innovation unless information is available about the extent to which the innovation has actually been implemented (Charters & Jones, 1975; Gephart, 1976; Hyman, Wright, & Tompkins, 1962). This has been a particularly vexing problem as planned innovations have become more complex; estimating whether implementation took place involves going beyond the simple question, "Was the innovation implemented—yes or no?" We need to be able to discern degrees and modes of implementation. Answers are needed to the questions of where, how, and how much implementation took place.

Researchers have begun to grapple with this issue and to explore the conceptual and measurement issues related to the crucial stage of implementation (Berman & McLaughlin, 1977; Greenwood *et al.*, 1975; Stebbins, Proper, & Ames, 1976; Vall, Bolas, & Kang, 1976). Several investigations which were being conducted concurrently with this study have been particularly enlightening in this regard. For instance, in a study of differentiated staffing, Charters and Jones (1975) question the emphasis placed on evaluating *outcomes* of change when insufficient attention has been paid to whether or not the intended changes actually occurred. They point out that such studies may well be appraising the effects of "nonevents." Charters and Jones identify four "levels of reality" of a school program: institutional commitment, structural context, role performance, and learning activities. They advocate the need to document actual program differences or changes, at least at the level of role performance, in order to verify that implementation took place. Only then can one confidently consider the outcomes of implementation.

The Rural ES Program in the Research Context

The study of the Rural ES program offered an array of opportunities to fill gaps in knowledge about the process of change. The emphasis was to be on organization, not individual, factors that aided or impeded change. The change process could be followed from the planning stage through implementation and beyond. Data could be gathered on numerous aspects of implementation in order to determine the degree to which the planned changes actually occurred.

Perhaps the most significant aspect of the Rural ES program was its size and complexity. Because school systems had rarely tried to change many activities at once, the Rural ES program could be seen as a pioneering attempt to provide insights into the possibilities of large-scale and comprehensive change in educational systems. Studying the implementation and outcomes of such comprehensive changes would in turn be a complex and challenging task.

Despite the potential difficulties involved, the time was ripe for a large-scale research effort. Indeed, the Rand Corporation's Study of Federal Programs Supporting Educational Change (sometimes referred to as the Rand Change Agent Study) has recently demonstrated that the study of complex as opposed to simple projects provides greater insight into the change process. Complex projects are defined as covering the entire student body or the entire classroom, large areas of curriculum, broad changes in teacher behavior, and integration of activities with other activities of the district's educational program (Greenwood *et al.*, 1975). Other researchers have also called for the generation of large projects that are intended to have major impacts on schooling by involving significant numbers of participants and requiring large investments of time, money, and personnel (Baldridge & Deal, 1975). If accompanied by systematic research, such projects can provide more useful insights into factors affecting the change process than can additional quantitative and qualitative studies of more limited change efforts.

The work presented in this volume addresses itself to a number of issues that can illuminate knowledge of the change process; these include: (1) the interaction between the external program sponsors and the school systems, particularly the development of this relationship and its impact on the change process; (2) the characteristics of the districts and of the communities in which they are embedded, and the impact of these factors on the implementation and continuation of innovative programs in the districts; (3) the relationship between districts and schools, and the impact of this relationship on organizational change; and (4) the characteristics of schools, and the impact of these characteristics on organizational change.

These issues are not novel; rather, they represent enduring questions for research and action in the field of organizational change. While previous studies have addressed themselves to these organizational constraints on innovative processes, rarely have these constraints been examined across a number of schools and districts in a systematic fashion. (See Chapter 2 for greater detail.)

RESEARCH THEMES AND ISSUES

At the outset of our study of the Rural ES program, we decided to take an exploratory approach. Research frameworks such as systems theory were used to organize the way in which data were collected and organized, but we attempted not to be overly bound by hypotheses developed at the outset of the research. We wished to allow issues to emerge and be explored as the data accumulated and initial analyses were conducted. Two major themes, however, have informed our research and the issues addressed in our analysis: the outcome of the change process, and the notion of system linkage.

Outcomes of the Change Process

Our approach to the conceptualization and measurement of outcomes was strongly influenced by the programmatic features of Experimental Schools (ES): local planning, comprehensiveness, and five-year terminal funding.

This stress on local initiative led to the creation of a series of very different local projects aimed at "comprehensive change," each specific to the circumstances of the district in which it was planned. The implications of this emphasis on the local plan for the study of change in schools were significant. As opposed to a situation in which several school districts and their schools attempted to implement a "treatment" (the usual case in social-intervention programs), the "innovation" being attempted differed from district to district and from school to school. This fact immediately raised a major methodological and theoretical issue that has plagued studies of organizational innovation for years: how to compare different phenomena occurring within different organizational contexts, all of which are assumed to fall into the same general class.

The mandate of comprehensiveness had several implications for our study of the process of change. First, it implied a scale of change that was relatively unusual in education, where problems are usually dealt with as they become particularly pressing and thus sequentially rather than simultaneously. The outcomes of the program at the district level intended by the federal planners were not very specific, but they were to occur in a broad range of functional areas. It would be difficult to measure and distinguish among the effects of this multitude of often vaguely defined program components.

Furthermore, the funds were authorized only to help the districts transform their educational programs, and the districts had to agree to assume any continuation of costs by 1977. The funds were not intended to subsidize activities previously under way or to pay for routine capital improvements and operating costs. The money could be described as a five-year "bubble" in the overall school-district budget to pay for "transition costs." At the end of the five-year period, each district was expected to be doing very different things educationally, either at about the same per-pupil costs as existed prior to the advent of their ES projects, or at an increased level that could be borne by the school district itself.

In other words, despite the high goals, the program did not intend to make major changes in the long-range funding patterns for rural schools, which, as noted, generally suffer from inadequate fiscal support when compared to other school systems. Rather, the intent was to provide a relatively limited "shot in the arm" that would allow the districts to utilize their own limited resources more effectively in the future. This program feature raises a number of research questions. How do districts deal with a "windfall" budgetary increase of up to 15%, particularly when they know that it is temporary? What happens to the innovative programming and effort when the funding is terminated? How do schools and communities view this type of "catalytic" assistance after the first flush of exuberance and later as the influx of new funding begins to fade?

In attempting to deal with these issues, we became convinced that a tremendous weakness in previous research on educational innovation was its inability to determine the nature and fact of innovation. The question of how to measure innovation is particularly perplexing in schools, where most of the new developments that emerge either from local change programs or from R&D efforts can be classified as "soft" (such as changes in methods of teaching or in school organization) rather than "hard" innovations (such as the installation of a computer or an educational television system). As Gross *et al.* (1971) have amply demonstrated, the presence and use of "soft" innovations are difficult to measure from the reserach perspective, since they are difficult to understand and subject to many permutations in the implementation process. Thus it is fruitless to talk about innovative processes or the factors that contribute to such innovations in the absence of relatively "believable" evidence that change did, in fact, occur and that the change corresponds, at least partially, to the formal description of what was intended.[3] Chapter 3 discusses the measures

[3] This is not to say that the student of educational change should not look for unintended outcomes of any innovation program. However, it is clear that we need to avoid the frequent problem, encountered by Charters and Jones (1975), of "labeling," in which activities are called by a name associated with a new educational practice, but, in fact, bear no resemblance to this practice.

we developed to detect these sometimes elusive kinds of evidence.

In attempting to define and measure the outcomes of the change process, we have striven to redress what appears to be a growing imbalance in the literature on educational innovation. As noted above, until recent years, few studies of educational change examined what happened to the innovation after the decision to adopt it was made (Berman & McLaughlin, 1974; Gross *et al.*, 1971). In recent years, on the other hand, there has been a swing of the pendulum to the point where some researchers have claimed that the study of the *implementation and routinization* of new programs is the only needed emphasis for theory and practice (Berman & McLaughlin, 1974).

Like most recent studies, our efforts have been spent more on the explication of the later stages of change than on the earlier ones. However, we have tried to show throughout our study how earlier events affect, or are related to, later decisions and actions.

Finally, we have attempted to grapple with the problem of unit of analysis by committing ourselves to an attempt to measure change at the organizational level rather than an individual actor's use of an innovation. While both individual change and organization change are important in a complete view of the process and outcomes of an effort to improve schools, we have chosen to design measures of change which do not rely on the aggregation of individual changes.

System Linkage

A second pervasive theme underlying our research is one which we call *system linkage.* Our premise is that educational organizations do not exist in isolation but instead are embedded in a complex social system and are in continuous interaction with their environment. For example, the school and the school district are both located in a larger external environment—a community of parents and other residents; and in a larger educational environment—educational organizations at the state and federal levels. The federal level is particularly important; federal involvement in local school systems frequently bypasses the state level to deal directly with local school districts.

The significance of this premise is reinforced if we apply an emergent concept in the literature on educational organizations—the concept of "loose coupling," as an explanation for some of the apparently random and unexpected behaviors of school systems (Weick, 1976).

The concept of loose coupling assumes that the parts of a school system are tied together tenuously and variably. For example, the central office may have varying degrees of influence over different schools in the same district. Its influence may vary for different decision-making or activity areas and may also fluctuate considerably over time. In sum, the concept of loose coupling attempts to describe some of the characteristics of complex school systems that are not accounted for by a formal organizational chart.[4]

One drawback of the term "loose coupling" is its implication that organizations fall into one of two categories, either "loose" or its opposite. However, organizations can more usefully be described as being located at different points along a continuum, whether on this variable or on any other. Hence, we use the term "system linkage," which does not imply a dichotomy, instead of "loose coupling." Furthermore, we wish to emphasize the concept of the school as a system of interrelated parts. Therefore, the inclusion of "system" in our term attempts to identify *what* it is that is tied together.

The degree to which parts of the school system are linked is of critical concern to this study because of the nature of the change intended within the ES program. Minor change occurs constantly in educational systems: individual teachers try out new approaches, principals alter the schedule slightly, or teachers are assigned to non-classroom activities. However, such changes do not require much cooperation among different parts of the educational system and have little system-wide impact.

The Rural ES program, however, envisioned *comprehensive change on a district-wide basis*. In other words, all parts of the system were to be affected simultaneously in a coordinated fashion. Clearly, this implied greater coordination among the parts of the system than typically occurs on a day-to-day basis in most districts. Thus, two significant questions concerning systems linkage arise: (1) Could districts muster the necessary internal coordination among system parts to plan and implement a change project that would meet the standards of the

[4]It is interesting to note that, with some regularity, organizational theorists emerge with a significant new concept which captures the irregularity of organizational behavior, and the ways in which it deviates from the constrained patterns implied by traditional administrative theory. Among the previous significant theoretical concepts which fall, along with "loose coupling" into this category are the concepts of "informal organization," "mock bureaucracy," "goal displacement," and "subordinate influence."

federal sponsors, and, if so, at what costs? (2) How did the degree of linkage among parts of the system affect the degree to which school districts were able to implement local ES projects?

The concept of system linkage seems to capture some of the most critical issues associated with the goals and objectives of the ES program and helps us formulate questions that are of great programmatic interest. Furthermore, the theme of system linkage has a broader application in that it facilitates and contributes to an emergent theory of change in educational organization. As will be shown in Chapter 2, the concept of system linkage is not only a powerful way of organizing some new ideas about the operation of schools and other organizations, but also a way of integrating existing empirical reserach about change.

We stress these two themes, outcomes of the change process and system linkage, in the chapters to follow. Specific issues within the ES program to which these themes apply are noted below as an overview of the organization of this volume.

OVERVIEW OF THIS VOLUME

The remainder of this volume is devoted to the story of the ES program and its effect on 10 participating rural school districts and their schools.

Chapter 2 elaborates on the research context of this study and addresses the objective of contributing to an emergent theory of organizational change in schools by relating existent research traditions to the system-linkage approach, and by showing how this approach addresses some controversies in the research literature. This chapter discusses the rational and nonrational aspects of the change process. Using the system framework as a guide, we outline some of the competing theories of change and describe ways in which linkage arrangements within the system may moderate the system's functioning.

In Chapter 3 we present an overview of the methodological approach to this study, including the research methods that were used, measurement problems that were encountered and how they were resolved, the data-collection procedures, and the nature of the data base that was used for the analyses described in this volume.

Chapter 4 introduces the federal and local partners in the ES change process. It details the explicit and implicit assumptions made by the federal planners of the ES program—assumptions concerning the nature of change, the design of the program, and the nature of the rural school districts. As we point out, federal assumptions did not always correspond to local assumptions, needs, or realities, and some of the federal expectations were mutually inconsistent. Next, the local school districts are introduced; we describe the 10 districts and discuss their apparent readiness for change.

Chapter 5 addresses the issue of federal-local relationships and the ways in which the nature and quality of the link between the federal and local levels affected the outcomes of the first stage of the ES program: the planning year and the final project plan that was required from the districts by the ES program in order to obtain long-term funding. A major emphasis of the chapter is on the negotiation of the ES program between its federal and local sponsors.

Chapter 6 shifts from the districts to schools within districts as the focus of examination. Despite the district-wide planning of the 10 ES projects, actual implementation occurred primarily at the school level. This chapter describes and analyzes implementation within schools and includes discussions of (1) how much was implemented, the differences in implementation across schools within districts, and the disjuncture between what was anticipated and what was actually implemented; and (2) what characteristics of the organization affected levels of implementation.[5]

Chapter 7 continues the analysis begun in Chapter 6, pursuing a

[5]The reader familiar with several of the companion volumes from the Rural ES study (Abt & Magidson, 1980; Clinton, 1979; Firestone, 1980; Herriott & Gross, 1979) may feel that there is some discrepancy between our story, which is one of modest optimism, and those of the other authors, which tend to stress the failures of ES. We believe that there are several important reasons for the differences in tone. First, our data cover a much longer span of time, and allowed for the real gains of the program to be better balanced against the turbulence and conflict of the early years of planning and implementation (see Chapters 4 and 5). Second, we believe that our comparative view across districts allowed us to find variation in implementation and change which on-site observers of single sites were unable to discern. Third, our emphasis upon looking for change in schools rather than in districts led, we believe, to the location of changes that were relatively invisible at the community- and district-office level. Many of these changes were worthwhile, even if they were not "comprehensive." Finally, there is a matter of perspective: We have attempted to make our work relevant to the development of proactive theory and policy regarding educational change, and, as a consequence, we have chosen to see our cup as half full rather than half empty.

more limited number of theoretical questions that emerge from the discussion of theories of change in Chapter 2. Particular emphasis is placed on the degree to which system linkage affects the implementation of change in schools. Three issues are highlighted: (1) the degree to which various groups of system characteristics interact and are analytically linked in explaining the outcomes of the change process; (2) the degree to which system linkage through the authority structure affects the process of change; and (3) the degree to which a group of variables that can be uniquely identified as reflecting "tight" coupling predict ES implementation.

Chapters 8 and 9 explore the dual themes of linkage and implementation in several ways. In order to do so, we return to a view of the change process at the district level which was the intended focus of the entire ES program. In Chapter 8 we discuss how much was implemented when change is viewed at the district level, the nature of the implemented innovations, and two sets of factors that are associated with district-level implementation: factors inherent in the program design, and the nature of the linkage between teachers and administrators. In Chapter 9 we address the question of how much continuation of project activities occurred at the end of the funding period. Although most components of the projects were discontinued in all 10 districts, either during the funding period or with the termination of funding, some activities continued, in whole or in part, and certain spin-off effects persisted. Some issues concerning continuation are addressed, such as the relationship between the scope of implementation and continuation, and the characteristics which typified high- or low-continuing districts. Although much of this chapter is devoted to a descriptive analysis of continuation, the ultimate emphasis is on the types of linkage between layers and levels of the school environment, and the outcomes of the change process, as well as on the relationship between the nature of the district's local plan and its ES outcomes.

The concluding chapter, Chapter 10, summarizes the findings, with an emphasis on the ways in which this study and the research themes it has developed can contribute to improved educational research, policy, and practice. The first part of the chapter focuses on a theoretical summary. Chapter 10 concludes with suggested implications for individuals at the federal and local level interested in the design and management of planned change.

2

The Research Context

The topic of organizational change has received increasing attention from organizational and administration theorists and researchers in recent years. Although there has been some interest in natural adaptation or modification of organizations in response to unplanned events (Grusky, 1961; Meyer, 1975; Terreberry, 1968), the greatest emphasis in applied research has been placed upon the processes, determinants, and outcomes of *planned* change. This concern has nowhere been more evident than in the area of education, where schools and school districts have been subjected to considerable internal and external pressures during the past 20 years to change their curricula, their structures, and even their goals. Recent syntheses of the literature in this area have covered the research findings in considerable detail (see, e.g., Fullan & Pomfret, 1977; Gaynor, 1975; Giacquinta, 1973; Pincus, 1974).

Different perspectives on organizational behavior have contributed to our own view of organizations, and discrepancies among them have suggested topics for us to consider. Our objective has not been to test a series of specific hypotheses about change in schools. Rather, we have taken on a broader mission of exploring the variety of data available through our study of the ES program in the expectation that it will shed some light upon several theoretical issues that have, we believe, inhibited the development of a coherent theory of organizational innovation.

There are two major approaches to the study of organizational change. The first focuses on change as a *rational,* manageable process which takes place as a result of active choices. From this perspective, the success of a program of change depends largely on the degree to which needs are appropriately assessed, plans generated, consensus developed among those who must make the changes, and resources accumulated to support the plan of action. The rational approach tends to emphasize change as a *process* and gives little attention to the organizational context in which change occurs.

The second approach views change from a *natural-systems* perspective, emphasizing the nonrational elements that condition the change process. From this perspective, change is not totally predictable, but is considered a negotiated process involving mutual adaptations of the innovation objectives and the context in which their implementation is being attempted. Factors less amenable to planning, such as organizational norms and structures and organization climate, are assumed to have at least as much impact on the success of a change program as the conditions stimulating the change or the rational plan.

Our basic assumption is that change in complex organizations, such as school systems, is mediated by both rational and nonrational aspects of organizational functioning (cf. Corwin, 1973; Gaynor, 1975; Thompson, 1967).The development of a theory of change requires integration of the two approaches to the study of change that presently exist. In order to contribute to such an integration, we have attempted to incorporate elements of both approaches.

ELEMENTS OF THE RATIONAL APPROACH TO CHANGE: THE STAGES OF CHANGE

Traditional organization theory maintains that change occurs as the result of rational choice (Cyert & March, 1963). According to this theory, the characteristics of an innovation that will make it most attractive include ease of explanation, possibility of partial trial, simplicity of use, congruence with existing structures and values, and obvious superiority to past practices (Rogers & Shoemaker, 1971).

At the individual level, the rational change process has been summarized by March and Olsen (1976) as a "complete cycle of choice": members of an organization observe various problems or needs within the organization; they analyze these needs in the context of their environment; they initiate new behaviors which affect organizational activities; the environment reacts to these innovations; and a new cycle of innovation begins. As Sieber (1972) points out, this approach views the innovator as a "rational man" who, when provided with clear information about an innovation that is applicable to his situation will, with little further stimulus, use it. At the organizational level, the view is much the same. Organizations and their leaders are involved in constant cycles of assessing organizational performance gaps, developing choices to improve the existing situations, implementing these choices, and observing internal and external reactions to new behaviors, procedures, or structures.

Many early organization theorists focused on the role of leadership in this process and the importance of the characteristics and style of the top administrator or set of administrators within the organization. Barnard (1938), for example, emphasized the fragile nature of organizational systems, and the role of the executive in ensuring that the "cooperative system" was maintained and functioned with relative smoothness. Selznick (1957) similarly emphasized the key role of leadership in inspiring subordinates, and in ensuring that goals and plans were shared and provided motivation. The role of leadership is, according to Selznick, particularly critical in times of instability or change, where the need for a clear vision at the top of the organization is greatest. The human-relations school, which was a dominant approach to organizational theory in the 1950s, emphasized leadership responsibilities in providing the structures, the incentives, and the interpersonal climate that would permit subordinates to obtain satisfaction from their work (Miles, 1975).[1]

[1]An alternative view of organizational behavior and the role of leadership had its origins in the work of Weber (1947), who emphasized the inherent instability of "inspirational" *leadership*, and focused on the notion of interchangeability of parts including the human occupants of positions within bureaucratic organizations. In such a system, it is rules and a formal *authority* structure that make the system work, and not any personal characteristics either of subordinates or of the chief administrator.

The Weberian view tends to predominate in organizational and management theory today, perhaps because of a general decline of faith in government and leadership. Whatever the cause, one can read through recent volumes of the *Administrative Science Quarterly* or other similar journals and find few references to administrative leadership or the functions of the executive. Recent articles that have drawn empirical attention to the importance of the elite in determining organizational behavior seem to have had relatively little impact on the overall course of organizational research (see, e.g., Hage & Dewar, 1973).

Characteristically, the rational approach breaks the change cycle into a series of stages, at each of which decision makers assess the state of the organization and make plans that will effectively thrust the organization into the next stage. The stages of change have been defined in a variety of ways by different authors. Some theorists have developed models of the innovation process based on individual choice (Klonglan & Coward, 1970; Rogers & Shoemaker, 1971), while others have developed stages based on the organization as the innovator (Berman & McLaughlin, 1975; Hage & Aiken, 1970; F. Mann & Neff, 1961; Wilson, 1966; Zaltman *et al.*, 1973).

A shortcoming of many of these models is that they emphasize the early stages in the process of innovation and to a certain degree neglect the later stages: implementation and institutionalization. Furthermore, they often ignore the potential aspects of change. Many attempts to implement previously agreed-on changes fail. Many educational innovations are not continued or institutionalized after they have been implemented.

Our research has incorporated the rational approach to change by using a relatively simple four-stage model of the change process similar to Hage and Aiken's. This model includes the important later stages in the process. (A summary of five different approaches to the stages in the process of organizational change is found in Table 1.)

The four stages which we found to be most relevant to and compatible with the ES approach to the process of change are readiness for change, initiation (planning for change), implementation, and continuation (persistence) of change. We do not wish to imply, however, that these stages occur in an orderly, linear sequence. In reality these stages overlap to a great extent; one stage often begins before the previous stage has ended. Nor are the timetable and intensity of activities at each stage necessarily uniform across all organizations even though they are engaged in similar programs of change. Each school district, by going its own way at its own speed, may spend more time at a particular stage or skip a stage entirely. Nevertheless, our four-stage model has been useful for both theoretical and programmatic reasons. One of the most important assumptions underlying the rational view of the change process is that successful innovation must be based on the adequate attainment of certain (generally unspecified) levels of success at each prior stage. This assumption has been revealed most tellingly in studies of failure to change, where

Table 1. Stages in the Process of Organizational Change

1. Evaluation	1. State of organization before change	I. Initiation	1. Conception of the change	1. Readiness
		1. Knowledge awareness substage		
	2. Recognition of need for change		2. Proposing of change	
2. Initiation	3. Planning change	2. Formation of attitudes toward the innovation substage	3. Adoption and implementation	2. Initiation
		3. Decision substage		
3. Implementation	4. Taking steps to make changes	II. Implementation		3. Implementation
		1. Initial implementation substage		
4. Routinization	5. Stabilizing change	2. Continued substained-implementation substage		4. Continuation
(Hage & Aiken, 1970)	(F. Mann & Neff, 1961)	(Zaltman *et al.*, 1973)	(Watson, 1967)	(Rosenblum & Louis, 1977)

failure is often accounted for by referring back to inadequate provisions of supportive resources (see, e.g., Gross *et al.*, 1971). Thus, successful planning may be difficult for the organization that has not exhibited signs of readiness for change. It will not have the internal resources (an awarness of its problems, experience in dealing with innovations, and so forth) that will assist it in developing appropriate and well-conceived strategies for making changes. Similarly, poorly conceived plans will be difficult to translate into activities associated with implementation. Finally, in order to achieve changes which persist over time, implementation must reach a certain level, or the system will slide back back to where it was before its effort to change.

The stage model was appropriate for inclusion in our framework for another reason. The concept of "stages" was implicit in Washington's design of the ES program. Districts were to be selected on the basis of a displayed readiness for change. They were allowed a year in which to assess their needs further and design a plan for comprehensive change consistent with their needs. A contract was signed through which the districts were authorized to implement their plan. (Chapter 5 discusses this process in great detail.) Finally, the districts were to include in their projects provisions for sustaining the changes beyond the federal funding period. Milestones that would mark the transitions from one stage to another included the signing of a contract at the end of the "planning year" and the renewal of the contract at the end of a three-year implementation period.

Although the separation of the various stages will not always be empirically distinct, division into stages helps us view change as a process rather than as a single action or event. The four stages are described below.

Readiness

Many theories define the first stage as one of the awareness, interest, or evaluation (see Table I). All 10 districts in this study had exhibited such activity by submitting a letter of interest to ES/ Washington. Even when information about oportunities for innovation and new resources is widely disseminated, organizations vary in their readiness to respond to such opportunities. The ES program distributed its Announcement to over 7,000 small school districts serving rural areas: fewer than 5% indicated a readiness to embark on a change process by submitting the required letter.

However, the degree of readiness of an organization for change is likely to vary and to be a function of a variety of factors. For example, certain antecedent conditions in an organization may serve as fertile ground for the seeds of change, such as attitudes supporting change, recognition of unmet needs, clarity of goal structures, organizational structures and leaders favoring innovation, professionalism of staff, and few strong vested interests in preserving the status quo (Glaser & Ross, 1971). Other conditions may include a history of successful change or pressure from the environment that is compatible with the organization's previous commitments (Corwin, 1973; Gross *et al.*, 1971; Perrow, 1974). Finally, uncommitted resources must be available to the organization: money, personnel, time, skill, and a tolerance for initial failure (Gross *et al.*, 1971).

Initiation

Initiation is marked by a period of intense planning during which a set of problems must be dealt with by those who wish to introduce the innovations. One problem is the need for a "change agent" from inside or outside the organization to facilitate the change by lending support, prestige, advice, and leadership. A major decision to be made is whether to involve subordinates in the process at this stage in order to achieve higher morale (Bennis, 1966) or greater commitment (Goodlad & Anderson, 1963). An underlying assumption of many organizational theorists is that if those who will be affected by decisions are involved in decision making, they will be more willing to make adjustment necessary to help implement change (Coch & French, 1971; Gaynor, 1975; Simon, 1965). It is generally maintained that wider participation of diverse groups will lower the probability that various needs will be overlooked, thereby enhancing implementation and overcoming resistance to change (Bennis, 1966; Coughlan & Zaltman, 1972; Havelock *et al.*, 1971). The evidence, however, is not clear-cut, and a number of authors have questioned the generalizability of the relationship between decentralization of decision making in planning and subsequent successful implementation of planned innovations (Adams, Kellogg, & Schroeder, 1976; Arnn & Strickland, 1975; Lischeron & Wall, 1975). In any case, the school districts participating in the Rural ES program had little choice in this matter. ES/Washington required that planning-year activities stress the collaboration of constituent groups in the school districts as well as with the federal monitors.

Implementation

Many of the problems and disappointments with large-scale innovation efforts (referred to in Chapter 1) have been attributed to the failure of planned innovations to actually be implemented. In recent years, the research literature has begun to examine previously overlooked issues and factors affecting implementation (Berman & McLaughlin, 1975; Charters & Pellegrin, 1972; Gross *et al.*, 1971) as well as conceptual and measurement issues concerning the characteristics and dimensions of implementation *per se* (Fullan & Pomfret, 1977; G. Hall, Loucks, Rutherford, & Newlove, 1975).

Ideally, successful implementation follows from careful initiation, but often the two are intertwined. If initiation is a planning phase, then implementation is a trial stage. It is that part of the process of planned change in which new procedures and activities identified during the initation stage move from ideas on paper or in the minds of planners to changes within the organization itself. It is in this stage that the new program becomes a reality to most members of the organization, and conflict may arise from contingencies that have not been foreseen (Hage & Aiken, 1970).

When the actual attempt to implement a change begins, the system must deal with important questions. Should those who are to participate in implementation also participate in decision making about the implementation process? How will the system mimimize internal resistance to innovation or change? How many existing programs must be altered or modified? To what degree must they be changed? Should the system establish a temporary subsystem to introduce the new method, practice, or policy? Gross *et al.* (1971) have described in detail these characteristics of the implementation stage in the introduction of a new teacher role model within an urban elementary school.

The system must also consider the impact of program change on the environment. In some instances the survival of the system in its sociocultural environment may be threatened by the improvement being implemented. Thus the risks are great and the responses to innovation are complex: small wonder that many innovations get lost or distorted in the transitions from initiation to implementation (Pincus, 1974). It is entirely possible that during the implementation stage the nature of the change itself will be modified. As Sarason (1971) points out, the adoption of an innovation, even if it is fully understood and well planned, does not guarantee that it will survive intact. He cites the example of new math, which often seems to end up being taught in the same way as old math.

The conceptualization and measurement of implementation continues to be a multifaced construct in search of an explicator (D. Mann, 1976). As Giacquinta (1973) suggests, the study of change appears to have been based on a truncated conception of the process. But researchers have begun to grapple with implementation issues and to attain further knowledge in this area. Fullan and Pomfret (1977), for example, have identified four categories of factors which affect the *degree* of implementation: characteristics of the innovation, including explicitness and complexity; strategies for implementations; characteristics of the implementing unit or situation; and macro sociopolitical factors.

G. Hall *et al.* (1975), stressing that assessment of implementation processes and inputs must take account of variations in level of use both across individual users and over time, have identified eight discrete levels of use. Emrick (1977) measured implementation outcomes in terms of scale of adoption, fidelity, nature of local adaptation, local attitudes and opinions regarding the innovation, and anticipated likelihood of continued use.

Our contribution to the conceptualization and measurement of implementation is the construct we call "scope of change," which includes both quantitative and qualitative dimensions. These are discussed in detail in Chapter 3 and in the analytic chapters which follow.

Continuation

The fourth stage in the process of change has not been clearly depicted in the research literature. It has been labeled "routinization" (Hage & Aiken, 1970), "continued sustained implementation" (Zaltman *et al.*, 1973), "institutionalization" (Milo, 1971), and "incorporation" (Berman & McLaughlin, 1975). Each of these terms connotes a somewhat different aspect of this process. There is, of course, a possible alternative to continuation, one which Rogers (1962) has labeled "discontinuance."

This final stage in the innovative process may be defined in terms of the persistence of changes or the long-term effects of a program of innovation. The innovative effort must become incorporated as an accepted part of the organization's functioning. It must no longer be viewed as "experimental," but as ongoing or "routine." This can occur gradually; changes may simply continue or persist. However, the term "incorporation" implies that changes are actively absorbed in the organizational system as standard characteristic features. This absorption is aided by routinization factors; that is, events that help make innovations a routine part of the organization. Hage and Aiken (1970) have identified several factors that indicate the routinization of changes: the codification of new rules and regulations associated with the innovation, detailed job descriptions for new roles associated with the innovation, the replacement of personnel who were centrally involved in their implementation, and the development of training programs for new replacements. Yin and Quick (1977) have expanded these notions of routinization to include the concept of "passages and cycles." An important passage is the shift of fiscal support from "soft" money to "hard" money (i.e., from external to local funds). Innovations are said to be routinized when they survive over several cycles—for example, annual budget reviews, turnover of staff, or training cycles. A project is institutionalized when the district's budget process, personnel allocation, support activities, instructional program, and facilities assignment routine provide for the maintenance of the innovation (Berman & McLaughlin, 1977).

Rational models of change assume that successfully implemented innovations will be continued. However, certain factors may militate against incorporation. For example, conflict over the innovation may make the cure seem worse than the disease. The system must also decide if the cost of continuing a change is worth the benefits it yields. Furthermore, since funding for implementing innovations often comes from the federal government, aspects of the federal-local monitoring process or of the funding itself may discourage incorporation. Pincus (1974) has pointed out several factors that may lead to discontinuance: (1) the tendency of the federal government to subsidize educational research and development without particular reference to its effects on various outcomes of schooling; (2) frequent changes in program priorities and too short a life for educational experiments; (3) the tendency of the federal government to view their contributions as seed money to be replaced by district funds, while school districts find the typical cost of such programs beyond their ability to finance; (4) the federal government's support of innovation on a relatively small scale compared to other programs such as impact aid and compensatory education.

The design of the ES program was intended to overcome some of these difficulties. However, as will be seen in Chapter 4 (and in subsequent chapters in which the outcomes of the ES process of change are discussed), the lack of clear federal objectives and the type of federal/local interactions which took place hindered the process as well.

ELEMENTS OF THE NATURAL-SYSTEMS APPROACH TO THE STUDY OF CHANGE: THE SYSTEMS FRAMEWORK

The accumulation of data on the results of change programs has led to considerable dissatisfaction with a simple rational model. Peterson (1977) has recently summarized some "anomalies" that cast doubt on the rational approach:

1. Innovations are seldom implemented as planned. Rather, they tend to undergo a process of continuous change as they enter the system. These changes result from unanticipated characteristics and events.

2. The introduction of identical innovations within outwardly similar organizations may lead to different implementation processes and outcomes.

3. Different implementation approaches and change-management strategies may produce similar results.

These anomalies suggest that rational-choice models cannot fully account for the outcomes of change programs. A recent study of federal programs to support educational innovation provides clear support for these criticisms (Berman & McLaughlin, 1977).

What are the alternatives to the rational approach? One alternative that has marked recent research on educational organizations stresses the dynamics of schools as *social systems* (Gross *et al.*, 1971; Sarason, 1971). School districts are a type of human-service organization. Like other such organizations, they are social units deliberately established to attain a variety of explicit and implicit goals. They operate in a rational mode to the extent that they are able to proceed toward the attainment of specified goals in a classic bureaucratic manner logically consistent with existent knowledge and expertise. However, political forces within the organization, such as desires on the part of its members for power or status, can interfere with its rational decision-making process and smooth operation.

It should also be stressed that educational organizations are not closed systems; rather, they are in constant interaction with other organizations (Corwin, 1972) and with their sociocultural environments (Buckley, 1967; Herriott & Hodgkins, 1973; D. Katz & Kahn, 1966). They are dependent on their environments for legitimization, source of funding, clients and staff, and other influences. As a result, interference with the rational workings of the organization may come from external forces which the organization has overlooked or which it cannot control, as well as from internal political forces.

The systems approach does not imply that change is irrational, but rather that it may be a negotiated process involving compromises between rational choices and system characteristics (Corwin, 1973). The systems approach reveals some of the elements that may condition the rational decision-making process associated with change. Schools that are very similar in their teacher, student, and community characteristics may still have very different structures; as a result, the implementation process and its outcomes will vary from school to school. The systems approach also reveals another reason why a purely rational model is unlikely to predict outcomes. Systems, particularly organizational systems, are complex: they combine too many district characteristics to allow them to be easily analyzed or holistically understood by their members. Further, if the rational decision maker were able to map all of the characteristics of the system, he or she would be hard pressed to decide how they should be analyzed to predict an optimal change strategy, given the lack of sophisticated information about how the elements of the system combine to affect change programs.

The contribution of the systems framework to our study of organizational innovation has been twofold. We accept the image of a dynamic, differentiated organizational structure that it presents. Second, it has helped us identify sets of variables that may condition the rational decision-making process. A complete list of such variables would reflect aspects of the system's environment, input (materials, personnel, and information), throughput (in this case, pupils), output (knowledge, skills, and orientations of pupils at the time they leave school), and the structure and shared culture of the organization (Herriott & Hodgkins, 1973). All these elements of a system are interrelated, both in the ongoing normal operations of the system and in any process of change within.

The systems approach to the study of organizational change and its implications for outcomes of change efforts has been characterized in two ways: one mode, that of conventional systems theory, purports that when one aspect of an organization changes, it affects others. For example, when interdependence increases, so does coordination. Increased individualized instruction creates more interdependence and specialization. Instructional individualization can create a need for school-level coordinators or district-level specialists. In sum, in organizations many things are linked. Changes in one aspect can change others.

On the other hand, the assumption of a tight interrelationship among system variables has been seriously questioned by those who have pointed out that systems are not always so tightly coupled (Corwin, 1977; Deal *et al.*, 1975; Meyer & Rowan, 1977; Weick, 1976). Rather, organizational subsystems are likely to vary in the degree to which they are intimately linked. Consequently, changes in one aspect of a system do not necessarily affect others—particularly in organizations which are highly ambiguous (March & Olsen, 1976). Changes in institutional patterns can occur without affecting roles, relationships, or norms. Changes in one part or level can be localized, or produce only symbolic alterations in others.

As schools begin to alter conventional patterns and practices, resistance, conflict, and turmoil are inevitable and may have unintended organizational outcomes. In the following sections we elaborate on how we have drawn from the variants of the systems approaches to organizational change that we have noted above, and on how we have selected variables from the various approaches or traditions.

Alternative Theories of Organizational Innovation within the Systems Framework

Despite their common opposition to rational theories of change, systems theorists diverge widely when it comes to explaining the outcomes of change programs or suggesting effective strategies for organizational change. Although all see organizations as interconnected systems, they give primacy to different parts of the system in explaining change. We may identify four alternative approaches, each of which stresses one of the system elements identified above: culture, structure, input (e.g., staff characteristics), and environment. Variables representing these four system elements are the ones we have chosen to emphasize in this volume.[2]

Theories of Culture and Change

Culture can be defined as the meaning collectively shared by members of a social system (Parsons, 1960). Although to some extent constrained by various structural arrangements, the culture of educational systems can be a potent, independent force in the process of organizational change (Sarason, 1971). The cultural approach emphasizes the importance of the people in an organization (Likert, 1967). In order to change the organization, one must first change the ways in which the people in it view their behavior, or the ways in which they relate to one another (Alderfer, 1971; Bennis, 1966; Pritchard & Karasick, 1973). If and when significant groups of people within the organization have orientations that support change programs and that are consistent with the objectives of the change programs, then change will (or can) take place (D. Katz & Kahn, 1966; Schein, 1969). Few proponents of this approach believe that structural variables are of total insignificance (indeed, many of the field experiments within the human-relations tradition also manipulate some structural aspects of the organization); however, the general emphasis is upon organizational culture and its impact on change.

[2]Though throughput and output are no less important, they are the major focus of other products of the research on the Rural ES program. See, for example, Abt and Magidson (1980).

Theories of Structure and Change

A second tradition derives from a more sociological perspective and focuses on the Weberian approach to organizations and their structure (Blau, 1972; Pennings, 1976; Perrow, 1972; Pugh, Hickson, Hinings, & Turner, 1968). Structure is defined as "the relatively enduring stable *pattern of social interaction* which integrates the various and sundry elements of a social system" (Herriott & Hodgkins, 1973, p. 26). Particularly important to organizations in general are their complexity (e.g., levels of authority, distinct occupational roles, subunits), authority structure (degree of participation of different members in decision making), and formalization (degree of codification of jobs) (Hage & Aiken, 1970). Also of importance are the variables of size, dispersion, technology, and autonomy (Price, 1972).

A notable aspect of the structure of organizations is their tendency to become increasingly differentiated and elaborated (D. Katz & Kahn, 1966). The process of differentiation produces a plurality of parts or subunits, each of which is devoted to achieving a limited set of objectives for the larger system (Argyris, 1959; March & Simon, 1958). Although these subunits are part of the larger system, they nevertheless have, by nature of their specialized functions, a certain level of autonomy, which inevitably produces tensions within the organization (Gouldner, 1959). It is, to a large extent, the proliferation of partially autonomous subunits, in addition to the uncertainty of relationships between the system and the environment, that makes a fully rational approach to organizational analysis impossible.

Writers in the structural tradition tend to view culture variables as outcomes of structural arrangements. Basically, the structural aspects of the organization (its complexity, formalization, authority structure, and so forth) are seen as constraints on individual behavior (Blau & R. Hall, Hass, & Johnson, 1967; Klatsky, 1970). In its extreme forms, the structural approach is an analogue of behaviorist therapy for individuals: it assumes that if the structure is changed, then changes in the informal organization will be a natural result. Indeed, proponents of this approach hold that the same individual will behave in different ways within different structural arrangements, whatever his or her informal orientations may be (Woodward, 1965). Thus, structuralists tend to assume that their preferred variables will provide the most comprehensive and complete explanations of naturally occurring change, to which cultural variables will contribute insignificantly.

Theories of Input and Change

A third approach to the explanation of organizational change is based on such input variables as staff or individual characteristics. The staff of the school are critical to the innovation process; they are the planners, adopters, and final evaluators of the attempt to change the system. For example, a major consideration in the assessment of possibilities for initiating change is the skill level of the teachers (Corwin, 1973). Whether they are principally from the local area, from nearby cities, or from other regions can make a difference in their attitude toward innovation; it may also make a difference in their ability to manage the innovations that are being implemented. Teachers who have worked for a long time in a static educational system may feel very uncomfortable with the prospect of systemic change. This approach is best exemplified in the diffusion literature, which emphasizes the characteristics of individual innovators as well as the social-interaction network in which they are embedded (Berelson & Steiner, 1964; Carlson, 1965; Havelock, Huber, & Zimmerman, 1969; Rogers & Shoemaker, 1971). The diffusion literature suggests that educational innovators are likely to be males, older, less satisfied with their careers, of higher social origin and education, and of greater cosmopolitanism as measured by travel, journals read, and work experience in a variety of contexts (Carlson, 1968).

The jump from the act of an individual adopting an innovation to the characteristics of an entire staff as predictors of organizational innovativeness is rather a large one. Nevertheless, several recent studies have attempted to examine staff characteristics and innovation (Corwin, 1973; Hage & Aiken, 1970), and have found such characteristics to be important predictors of change.

The diffusion perspective is bolstered by other theories that emphasize the characteristics of the staff. Corwin (1973) identifies what he terms the "replacement" approach to change, which emphasizes the importance of developing—often through recruitment—both leaders and powerful subordinate groups with characteristics that will facilitate the change process. The replacement theory emphasizes the significant power of groups of resisters among the staff to undermine innovations that they disapprove of. (See, for example, Barnard, 1938, and Mechanic, 1962, for discussions of the power of subordinates over organizational behaviors.) As a consequence, replacement theory argues that one of the most effecfive ways of facilitating change is to staff the organization with people who have appropriate demographic characteristics and supportive attitudes (Clark, 1960; Guest, 1962).

Theories of Environment and Change

A fourth approach to organizational change emphasizes the importance in the change process of environmental factors, such as other organizations or the larger social context. A number of authors contend that environmental factors increasingly determine organizational behavior in a modern society where the environment is rapidly changing (Lawrence & Lorsch, 1967; Terreberry, 1968).

The distinction between a system and its environment is never clear-cut. Although there are many ways to distinguish between educational systems and their environments (Griffiths, 1967; Owens, 1970; Siegal, 1955), a recent study (Herriott & Hodgkins, 1973) seems particularly useful as regards organizational change in school districts. There, the authors conceptualize the American educational system at five distinct *levels*—school, school district, state educational system, regional educational system, and national educational system; and the system's sociocultural environments at five corresponding *layers*—neighborhood, community, state, region, and society. Thus, changes occurring in a school district will be affected by its sociocultural environment, the community layer. But they will also be affected by higher *levels* of the American educational system, which also are a part of the school district's environment.

The way in which external factors influence the innovation process has been variously assessed by different researchers. At one extreme, the environment is considered a basic, direct cause of organizational change. Bowles and Gintis (1972) and M. Katz (1971), for example, assume that schools are impervious to change except through basic, large-scale social reform; these writers view schools as vehicles through which the most powerful groups in society maintain their control. From another perspective, the environment is largely a stimulus. Baldridge (1971) argues that much of the change within any organization comes about in response to changing inputs from the environment (e.g., feedback on information). Still others appear to view the environment as a constraint. Wayland (1964), Herriott and Hodgkins (1973), and Oettinger and Marks (1974) assume that organizational change cannot take place unless it is compatible with the values and orientations of other segments of society (i.e., environmental layers) which are or may be affected. Herriott and Hodgkins (1973), for example, argue that the greatest changes in schools in less modern areas are likely to come "not from the local, state or federal initiative focussed directly upon the schools, but rather from external forces that can modify the socio-cultural context in which these schools exist" (p. 163). Finally, the environment may be perceived as crucial because it supplies resources. Corwin (1972) suggests, in this vein, that an organization is more open to change when it is able to cooperate closely with cosmopolitan organizations that can supplement its skills and resources.

The basic problem with these four theoretical approaches is that they are rarely examined in unison. The majority of empirical studies include variables that repressent only one or two of the system elements. For example, Clark (1972), Bowers (1973), and Bennis, (1966) focus primarily on variables that represent a cultural approach to change, while Deal (1975) and Hage and Aiken (1970) tend to emphasize structural features. Many studies, of course, include variables from several elements, but typical measures of theoretical constructs from one approach are more precise than those of competing approaches. More importantly, relatively few empirical studies of innovation actually test alternative theories that can be derived from the open-systems framework.[3]

This lack of integration in empirical studies is often coupled with a theoretical championing of one approach over another, with less attention given to the ways in which different system elements may be interrelated. Blau (1972), for example, stresses the importance of changing structures while neglecting the culture in which those structures are embedded. Argyris (1972), in contrast, emphasizes changes in culture but is apparently oblivious to the influences of structure.

[3]There are exceptions to this generalization. Baldridge and Burnham (1975) examine the relative contribution of leader characteristics, structure, and environment to innovation in schools. Hage and Dewar (1973) test the relative importance of values and structure as they relate to change. Corwin (1973) examines competing theoretical approaches to change, several of which parallel the traditions presented above.

Meanwhile, there is increasing empirical evidence that many factors may contribute to change and innovation in organizations of all types. Contingency theory represents one approach for dealing with the conflicting evidence from empirical studies. This approach maintains that relationships among given organizational variables depend upon the strength of other critical organizational characteristics (Friedlander & Brown, 1974; Litwak & Rothman, 1977; Perrow, 1972). However, contingency theorists have not yet attempted to use this approach to integrate competing theories from different disciplines or traditions.

This volume does not pretend to offer definitive results that will produce a final theoretical synthesis. However, we do examine the relative explanatory importance of the different system elements, both within the participating schools (Chapters 6 and 7) and at the district level (Chapters 8 and 9). From this examination, we are able to shed some light on ways in which the system elements relate to each other, and on ways in which they condition change and innovation processes in schools and school districts.

SYSTEM LINKAGE AND THE SYSTEMS FRAMEWORK

The systems framework assumes that change in any part of the system will have an impact that reverberates throughout the system. To take an example from recent organization theory, Hackman and Lawler (1971) assume that if a manufacturing plant initiates a job-redesign program, the new structures will affect the culture (morale) of the work place, the ability of the plant to determine input (recruit workers and reduce absenteeism), and the quality of the product (output). However, as we have mentioned, the assumption of tight linkage among system variables both within and between organizational subsystems may be challenged, particularly in the case of the typical American school district. The educational system is composed of classrooms, which are bound together in higher level organizational units (schools). Schools themselves are grouped into school districts, school districts into state educational systems, and state educational systems into regional accrediting agencies and into a national educational system. Due to long-standing traditions of local and professional autonomy, linkages become much looser as one moves from the level of the classroom upward. At the national level, for example, the "educational system" is largely informal and voluntaristic.

The degree of linkage has an important effect on the extent to which change in one part of the system will cause change in another. Thus, for example, if we examine the state level, we know that changes that take place within some districts generally have no immediate impact upon other districts in the state, and frequently have no long-range impact either. Within schools, on the other hand, a change in educational programs in the third grade can have a rapid and noticeable impact on the fourth grade. As the fourth-grade teacher receives incoming students who have a different background and level of preparation, she must adapt her own curriculum.

Deal *et al.* (1975) have noted that the "double segmentation" of schools within districts and classrooms within schools has led to situations where the variables associated with change at each level do not appear to be rationally integrated. In fact, others familiar with school systems have noted that one of their basic characteristics is the high degree of functional autonomy of parts such that even upheavals in the administrative central office may have little impact at lower levels in the system (Lortie, 1975). In some cases, there may be deliberate attempts to insulate parts of the organization from undue influence either from other internal units or levels, or from the environment (Thompson, 1967).

Recently, organizational researchers have eagerly taken up the concept of system linkage (generally referred to as "loose coupling"). It is increasingly clear that organizational and administrative theory is moving in a new direction—away from the traditional emphasis on system interdependence and toward emphasizing nonrational behavior and other deviations of social systems from the biological-systems model. For example, the concept of loose coupling is consistent with the growing interest in the explication of theories of decision making not based upon the notion of organizational rationality (March & Olsen, 1976).

The emergence of the loose-coupling or system-linkage approach should not be seen as a contradiction of previous systems theory. Rather, it is best viewed as an important extension of existing approaches to the analysis of organizational behavior from an open-systems perspective. Emphasizing variability in system linkage or interdependence of parts makes existing organizational theories much more congruent with observed organizational behavior.

Loose coupling is not, however, an entirely novel concept; early organizational theory emphasized not only the interdependence of parts within organizations, but also the ways in which the existence of multiple parts created problems of coordination, goal displacement, and conflict between the formal and informal organization. Corwin's (1977) review of the antecedents of the recent emphasis on structural looseness identifies several areas of concern that have a long history in organizational theory.

First, the nature of *control systems* in organizations has been viewed as problematic by many organizational theorists, who have emphasized the power of subordinates to evade rules and influence from higher levels (Barnard, 1938; Mechanic, 1962; Selznick, 1957). Second, many organizational studies emphasize the tendency of organizations to develop an increasingly complex *division of labor* which inevitably presents problems of control. The division of labor often occurs because of the need to develop some independence between task units—a need for what Gouldner (1959) refers to as "functional autonomy." Thus, the strain between autonomy and control is inevitable in a complex organization with a differentiated task structure. Third, the significance of *employer initiative* emerges in a number of organizational studies. The evasion of formal authority structures (e.g., rules) has often been noted to be functional, not only for the individual employee, but for organizational adaptation (Anderson, 1966; Blau, 1972; Corwin, 1965).

Although the construct of system linkage has theoretical antecedents, its value has to a large extent been ignored until recently. While there have been references in past literature to functional aspects of autonomy or rule evasion, the predominant orientation has been to view these characteristics of systems as "problems" that need to be overcome, either through the institution of close structural controls or through the development of greater participation and, presumably, stronger normative acceptance of organizational goals and procedures (see, e.g., Thompson, 1967, or Barnard, 1938). The recent emphasis on loose linkage is, on the other hand, associated with a strong conviction that it may not be bad for organizational health, and may in fact have some benefits. Weick (1976), for example, identifies the functional aspects of what he calls "loose coupling":

1. Loose coupling facilitates organizational survival by fostering stability. If organizations were required to respond to small changes in the environment on a regular basis, their energy for other productive activities would be lowered.
2. Loose coupling may facilitate organizational information gathering or "sensing," because different parts are all potentially able to respond to the same inputs (or feedback).
3. Adaptation or innovation may be facilitated in such systems. The possibility of innovation without disturbing the whole system allows the system to be more adventurous (or less restrictive), and the probability of innovations emerging is therefore higher.
4. Loose coupling facilitates local adaptation to unique problems and does not require standardized procedures.
5. System breakdowns or crises can be isolated so that deterioration of the entire system is prevented.
6. There is more room for self-determination of parts, an aspect which is consistent with contemporary value systems (and some psychological theories which see a sense of personal efficacy as a cause of better mental health).
7. Loose linkages make a system inexpensive to run, since they diminish the resources necessary for coordinating administration, conflict resolutions, and so forth.

Each of these functional characteristics of the more loosely linked system has its dysfunctional converse, also noted by Weick. However, on balance, the tendency of theorists of this persuasion is to look for evidence that highly controlling, tightly coordinated bureaucratic models are not necessarily more effective in many settings.

Definitions of System Linkage

Because the notion of system linkage has only recently assumed importance, much of the latest empirical and theoretical work has been devoted to explicating the utility of the concept for describing or explaining puzzling aspects of school behavior (Deal *et al.*, 1975; Meyer & Rowan, 1977; Weick, 1976). With the exception of Weick's listing of a range of phenomena that could be defined as evidence for loose coupling, little attempt has been made to systematically address the various definitions of the concept or to arrive at a potentially standardized vocabulary or set of indicators for the researcher.

In this volume we address the need for a definition of linkage that is sufficiently narrow to satisfactorily explain observations that emerged in our own and previous studies while not encompassing all organizational attributes and behavior. We look at two important kinds of linkage: that within a particular level of the educational system, and that between different levels.

Linkages within a Level of a System

The individuals composing any given level of the educational system may be viewed as units that can be more or less loosely linked. This issue is particularly salient for schools, where the linkage of individuals (teachers) is synonymous with the linkage of the basic organizational units—that is, classrooms. Intralevel linkage may be of three types. There may be *structural linkage,* or mechanisms which emphasize the formal means by which coordination of behavior is produced; there may be *cultural* or *normative linkage,* or mechanisms which emphasize the creation or coordination of similar behavior patterns through the development of shared definitions; and third, linkage may be characterized by *consensus between administrations and teachers,* which may have implications for the development of improved structural and cultural linkage.

Within a school system as in any other organization, intralevel linkage is a function of the degree to which there is coordination among potentially autonomous actors. Within formal organizations there are a variety of traditional methods for ensuring that activities are coordinated: the exercise of superordinate authority to mandate or control decisions about what activities are acceptable, the development of a system of rules which make explicit what people should (and should not) be doing, and the development of procedures for ensuring that desired behavior takes place and that undesirable behavior is kept to an acceptable minimum. In most organizations, these control procedures manifest themselves through rule enforcement and through methods for evaluating employee performance.

In a given organization, strong indicators of coordinated structure and high levels of administrative authority do not necessarily imply low levels of staff professionalization and participation. Rather, the level of influence or authority in an organization may be considered as a *variable* rather than a fixed sum. Increasing teacher or other staff participation in decision making will not necessarily reduce coordination, especially in the presence of strong administrative influence (Tannenbaum, 1960). The same cannot be said, however, for the concept of autonomy, which implies that the individual need not answer for his or her decisions to others. Autonomy is the very essence of loose coupling, for it implies that little or no coordination in decision making about behavior is required.

Linkages within a level are not confined to structural mechanisms for ensuring coordination. For example, nonbureaucratic organizations tend to deemphasize the use of rules and formal coordination mechanisms to ensure similarity in behavior between individuals (Litwak, 1961), and tend to emphasize *communication, collegiablity,* and *consensus building* as a means of achieving linkages between individuals or units. Such organizations are not necessarily placid. The maintenance of linkage through the building of high levels of normative commitment and consensus is not an easy task; differences of opinion between individuals or units must be addressed as soon as they arise. Although such "normative" linkage depends on consensus (Etzioni, 1964), the means by which consensus is reached is often through the resolution of disputes between members. Unresolved tension is anathema to the group whose linkage is based on consensus, and mechanisms for confronting disagreements are of utmost importance.

A particular type of consensus is critical in an organization that is a hybrid between the professional/collegial model and the bureaucratic model (as are most schools, according to Lortie, 1975, and Corwin, 1977, among others). We are referring to the degree to which consensus exists between the different actor groups—administrators and teachers—within a unit (school or district). Where consensus between administrators and teachers is high, the degree of system linkage is considerably higher than where the teachers' and the administrators' views of the organization are very disparate. Degrees of consensus may be of particular significance when one is assessing the quality of the educational system and its objectives.

Linkages between Levels or Units

Of equal importance as intralevel linkage is linkage between different levels of an organization (Bidwell, 1965). One aspect of this issue in educational organizations is the degree to which influence from the central office is conditioned by characteristics of the schools (Deal *et al.*, 1975). The concept of coupling has also been used to explain administrative effects on school- and classroom-level innovation.

To a large extent, linkage between levels is similar to that within a level, except that schools instead of teachers are being coordinated. However, some aspects of interlevel linkage are unique. For instance, the factor of spatial dispersion should not be overlooked (R. Hall, 1972; Louis & Sieber, 1979). Studies that have examined the effects of spatial dispersion have found that it has enormous impacts upon coordination of activities; the effort required to coordinate or influence dispersed units from a central location rises steeply with distance (Louis & Sieber, 1979).

Another factor affecting interunit linkage is the *number of units* that need to be coordinated. Although coordinating people is not an easy task, coordinating organizational units is even more difficult, and the larger the number of units that must be coordinated, the more difficult the task.

Spatial dispertion and number of units are both aspects of a particular problem of linkage which we call *physical linkage*. In addition to physical linkage between units, there is also the question of the degree to which units actually resemble one another. In other words, are the units standardized, like the typical fast-food franchise, or are they unique? The typical school may look very different from another school within the same district. Homogeneity of units may be a prerequisite for actual coordination; as any educator is quick to point out, developing similar behaviors or programs in schools that are markedly different from one another poses serious problems. Thus, from the point of view of linkage theory, homogeneity and physical linkage may be viewed as indirect measures of actual linkage between units.

It is important to emphasize that the causes of homogeneity or similarity between units within a system may vary widely, from the existence of a very homogeneous and significant immediate environment to the deliberate intervention and consistent monitoring that make McDonald's hamburgers taste the same from coast to coast. Within school systems, however, it seems to us unlikely that there would be great similarity between schools in the absence of some specific policies to encourage similarity. This assumption is made because we know that variability is a burning issue for central office personnel. They feel that variability is one of the greatest barriers that they must deal with; they see it as a phenomenon that arises spontaneously no matter what policies are promulgated by the district.

The Implications of System Linkage for Our Research

The notion of system linkage has several important implications for our study of the ES program. First, we assume that the ES program represented an effort to introduce system-wide changes within a school district; however, much of the focus of implementation must be at the school level. This is true because it is the schools rather than the districts that are vested with the responsibility for the day-to-day tasks of educating students. Second, we assume that both districts and schools may vary in their degree of linkage. Some districts may encourage schools to function with almost complete autonomy, while others have developed a structure and culture that encourages greater communication and linkage among schools.

These two assumptions imply that it is necessary to look at implementation at two organizational levels, at minimum. Because the programs of change were funded through the districts and were managed at the district level, we can expect that implementation characteristics will vary among districts. Since districts themselves may vary in their capacity to influence changes at the school level, it is also essential to look at school factors that may affect program outcomes.

In addition, the concept of variable system linkage stimulates a variety of significant empirical questions. For example, does tight linkage appear to facilitate or impede the implementation of planned change programs? Do different types of linkage have different or similar relationships with implementation and system change? Does the level of linkage change as a result of participation in a district-wide change program? Further, are the activities associated with the stages of dynamics of change? These and other questions related to the notion of system linkage are explored in the following chapters.

One deviation that we have made from many existing descriptions of schools from the loose-coupling perspective should be noted. Although many writers in this tradition concede that system linkage is a variable quantity, they ususally describe school systems categorically as being structurally loose, particularly with respect to instruction. (See, e.g., Meyer & Rowan, 1977.) We differ from this perspective in two ways. While agreeing that schools fail, in general, to conform to an ideal Weberian model of bureaucratic behavior, we do not necessarily assume that they are any more deviant (or loose) than other organizations of similar complexity. Rather, we agree with Corwin (1977) when he wonders why anyone ever really thought that organizations behaved according to an ideal type. Second, we assume that even if schools are, in fact, less tightly linked than other types of organizational systems, considerable variation exists both among schools and among districts in the degree of system linkage and the emphasis that is placed upon it. Thus it makes perfect sense to examine the impact of linkage even within a category of organizations that is more loosely linked than others. Indeed, it can do much to illuminate the problems of system linkage in general.

3

Methodology and Research Procedures

In this chapter we present an overview of the design and execution of our research, singling out those features of the program and research context of our study, of our general approach to the collection and use of data, and of our strategies for measurement and analysis that we believe are unusual or innovative. It is our intent to provide a description not only of these features, but also of problems and solutions that arose in the course of our research.

THE CONTEXT OF THE STUDY

This study of organizational change in rural schools is situated in two important contexts: the design of the Rural ES program, and the nature of the larger research effort of which this study was a part. Both of these contexts have significantly influenced the research endeavor.

The Program Design Context

The Rural ES program was designed as a series of "field experiments in educational change." The funding agency designed the program not as a pilot project that might continue indefinitely if proven successful, but as a natural laboratory for research. Unlike many studies of federal or state program interventions, the research was not an afterthought, but rather an integral part of the timing and conduct of the program. Thus, action and research went hand in hand. Research was to begin when the program started and to continue after the ending of the "action" in the local districts. In addition, all districts that participated were aware, from the beginning of their involvement, that they were to be "studied." Although some districts never became fully comfortable with the underlying philosophy of action research, all the districts cooperated with the research, to a greater or lesser degree.

An important feature of the program design was that it both mandated and supported a longitudinal study of organizational change. The objective of the research was to understand the dynamics and the outcomes of programmatic activities, and it was assumed that our approach would take full advantage of the opportunity to observe and measure the change process over five years, starting from its early stages. As it turned out, the research activities did not get well under way until late in the initial planning year. This late start had the doubly unfortunate effect of preventing direct observation during the earliest phases of the project, when crucial decisions were made, and making impossible collection of baseline measures prior to early implementation. It also necessitated very quick "start-up" of the research in order not to further postpone data collection, thus preventing the research team from getting fully grounded in the realities of the program during the design phase of the study. Nevertheless, since our work extended until nine months after the final funding of any activities in seven of the 10 districts, we were essentially an integral part of the action from its beginnings through the early phases of incorporation after termination of federal funding.

Although research formed an integral part of the Rural ES program, it was decided to keep research and action strictly separate at the operational level. As a result of this separation, researchers had to study the program as it evolved, with very little opportunity to influence the program in ways that might facilitate the actual conduct of the research.[1] To help keep activities at the local level from being influ-

[1]As noted in Corwin (1977), there was much ambiguity over this issue on the part of federal program officers, most of whom seemed to feel that the program was unduly

enced in any direct fashion by the ongoing research, the research in the districts was made as unobtrusive as possible. Districts did not receive any feedback about the results of surveys or other data-collection efforts until very late in the project. The amount of formal contact between the research staff and the local school staffs was minimzed, except that a researcher, who attempted to keep a very low profile, set up residence at each site. (The role of this "on-site researcher" is discussed in greater detail below.)

The Research Project Context

A second feature of the context of this study is that it was part of a much larger research project whose objective was to document and evaluate the functioning of the entire Rural ES program. The larger study, carried out by Abt Associates Inc. consisted of five major components. Three were studies of all 10 ES project sites; these cross-site studies called for uniform research designs across all 10 school districts and were directed by the research staff at Abt Associates' central office in Cambridge, Massachusetts. The two other components were in-depth investigations on an individual-site basis.

One of the cross-site research efforts was this study of organizational change in rural schools. A second major effort was designed to determine whether the innovations introduced by the 10 districts had any impact on pupils. That study focused not only on pupil achievement scores but also on other pupil outcomes, such as self-esteem, career goals, and satisfaction with school. The third study looked at the changes that occurred in the 10 communities with ES projects and the responses of the 10 communities to the activities sponsored by ES.

Of the site-specific research components, one consisted of a site history of each of the 10 communities and their school systems prior to

---

constrained by the preferences of the researchers. However, we often felt, as researchers, that we were unduly constrained by preferences of the program monitors.

The original scope-of-work statement for the overall project called for formative evaluation and technical assistance on matters of local evaluation, as well as for the types of summative research reported in this volume, but the inherent tension among the three proved, in this study of 10 geographically dispersed rural school districts, to be a particular problem very early in the project and led to emphasis exclusively upon summative research.

their involvement in the ES program. The remaining component of the overall research project was a series of case studies, which are holistic accounts of the change process encompassing diverse aspects of the community, schools, and individuals at each site. The case studies and the site histories, one from each of the 10 districts, were written by researchers—sociologists and anthropologists—each of whom lived in a single community for three years during the course of the project. These on-site researchers (OSRs) were selected, in part, because of their previous experiences in the conduct of ethnographic research. In general, they played traditional field-worker roles in their data collection and living patterns. In addition to collecting data for their case studies, in which they were allowed broad freedom to define their own research frameworks and strategies, they were available to assist other staff members in Cambridge in collecting data that would primarily serve the three cross-site studies.

Each of these five research components has produced a series of formal reports and other publications. (See Chapter 1, Footnote 2 for details.)

The project was designed on the basis of several important assumptions about the most effective way of answering overall study questions.[2] The first assumption was that each of the five component studies within the larger study had specific research objectives distinct from those of the other component studies. This, in turn, implied different research frameworks, instrumentation, and analytic objectives.

Second, it was assumed that enabling each component study to meet its research objectives required emphasizing its status as an equal partner in the objectives of the larger project and protecting its autonomy. Thus, for example, although it was expected that the OSRs would facilitate the studies of community, pupil, and organizational change through the provision of specifically requested information at periodic intervals, their major data-collection activities, which centered around producing the 10 case studies, would serve the other

[2] The overall study questions are addressed in Herriott (1980), which synthesizes the findings of the five separate studies. The questions are: (1) What social, political, and historical phenomena characterize the experimental schools? (2) What has been the impact of this program on pupils, schools, and their communities? (3) What changes persist beyond the period of federal funding? (4) What knowledge has been gained through this program for educational policy makers and practitioners?

studies only in ways that were consistent with their own case-study objectives. Similarly, teacher/administrator surveys that were designed by the organizational-change study might include some additional items to serve the needs of the pupil-change study, but these added questions would not have major impacts upon the design of the instrument. It was assumed that each study would serve the other studies to the degree that there was overlap in research questions, variables, or shared theoretical frameworks. By reading one another's work and through formal and informal contacts, senior research staff would share information with colleagues from the other parts of the larger research project.

The strengths of this overall design are great and deserve comment, particularly the degree to which they facilitated the analysis and writing of this volume. The autonomy of each of the component studies was a valuable feature, for, parallel with the assumptions underlying system-linkage theory, it tended to promote intellectual development and the emergence of new perspectives within each part. Since the research issues facing each study were differently timed, the intellectual development of each component occurred on quite a different schedule. If each study had been expected to move in similar directions at the same time, it is unlikely that the conceptual refinement and intellectual vigor of the studies would have been as great.

Second, by encouraging cross-fertilization during the course of the project, it became possible to integrate analysis across studies toward the end of the research period. This study, for example, has drawn heavily upon the work of the case-study writers to enrich our data base. The pupil-change study, on the other hand, has made considerable use of the quantitative measures of organizational characteristics and organizational change used by our own study, as has the community-change study. Integration was enhanced by the contacts that occurred over the life of the study, since we generally knew what the content and outcomes of analysis were prior to receiving written reports that could serve as actual data sources.

On the other hand, the research strategy of differentiating among the component studies during the collection and design while encouraging their integration in the final stages of analysis also has certain weaknesses from our perspective. For example, we came to view the case studies as an increasingly important source of data for both validating and interpreting our own results, and suggesting new analyses that might be useful. Our ability to draw upon the case studies directly was limited to the cases where they specifically used variables or addressed issues that were important to us. Since they were not written using our research perspectives, this often meant that we could use them primarily as illustration, rather than as a systematic data source. Similarly, our original intent had been to utilize measures of pupil culture obtained from the pupil-change study in our analysis, but the theory that directed the scaling activities within that study did not turn out to reflect our needs. Clearly if the entire project had been driven by the questions and approach that motivated the examination of organizational change in rural schools, our analysis might have been somewhat enriched. However, we must grudgingly admit that, had our perspectives been imposed upon the other study components, they would not have been able to achieve their own objectives as effectively.

This discussion is not meant to suggest that the study falls short of any reasonable expectations. In fact, given the size, complexity, and dynamics of a major research endeavor such as this one, and the original objective to conduct five separate (albeit overlapping) studies, the degree to which we have all learned from one another is very great. What is equally important is that we were able to draw upon perspectives that were broader than what was achieveable in our own work, and we could on many occasions supplement gaps in our own data with data available from other parts of the overall research project.

GENERAL APPROACH TO DATA COLLECTION

Our approach to data collection has been guided by the notion of triangulation (Webb, Campbell, Schwartz, & Sechrest, 1966), which emphasizes that multiple methodologies and multiple observations are better than single methodological approaches or unique observations. Our approach was based on two principles: that the *locus of the observer* of a social phenomenon is an important methodological variable, and that *both quantitative and qualitative data* should be used in studying the change process.

In studying social systems, data are potentially available from two different types of informants: from people who are actually *members* of the system, or from nonmembers who are observing the system. Both types of informants have value in social-science research, and each is the most appropriate data source for different data needs. For example, if one seeks to collect data on the characteristics of the system as they are *perceived* by its members, direct data collection from the members themselves is clearly quite appropriate. In many cases, however, members of a system are not necessarily the most reliable informants. As any field researcher knows, there is considerable variation among system members in the degree to which they have access to information about their own system, the degree to which they can objectively assess their system, and the degree to which they will willingly provide what information they possess to outsiders. In addition, system members may simply not see phenomena or behaviors that are of interest to the researcher, not because they do not exist, but because they are not recognized. (For example, a citizen of a tropical country is possibly a less reliable informant on the ways in which social patterns are affected by extreme heat than an observer who comes from a temperate climate.)

Furthermore, having information from system members *and* system observers is useful in determining the reliability of information provided by either source. If both outsiders and insiders agree that a phenomenon characterizes a system, one can place greater faith in the actual existence of that system property than if the two disagree.[3]

Even though there is continuing debate among social scientists about the relative importance and role to be played by qualitative and quantitative data, few would disagree that each provides a useful and different view of social structure and behavior. We do not wish to deal extensively with the arguments put forth by those who favor one form of data over another. Rather, our approach is based on the assumption that there are persuasive arguments for combining qualitative and quantitative data within a study, rather than relying exclusively on one form or another (Sieber, 1973).

Traditionally, studies of organizations have tended to rely heavily on methodologies involving the study of a single case (e.g., Blau, 1955; Clark, 1960; Gross *et al.*, 1971) or a comparative study of a limited

[3]We should emphasize that we realize that there is no hard and fast distinction between insiders and outsiders. The greater the amount of contact that an outsider has with a system, the more likely it is that she or he will reflect actual member views as opposed to partially analyzed views. Similarly, the role of the "marginal man" as an informant is well known among anthropologists and field researchers.

Table 2. Data Sources by Informant Locus and Data Type

	Data type	
Informant locus	Structured (quantitative)	Unstructured (qualitative)
System member	Professional personnel census forms	Telephone interviews
System observer	On-site researcher questionnaires	Case studies Site histories Special reports

number of cases (e.g., T. Burns & Stalker, 1961; Lawrence & Lorsch, 1967; March & Olsen, 1976).

More recently, the research emphasis has been upon gathering structured quantitative data across many organizations (e.g., Corwin, 1973; Hage & Aiken, 1970; Herriott & Hodgkins, 1973). In the past few years, researchers have attempted further innovations in methodologies, including quantitative secondary analysis of existing organizational case studies. (Dunn & Swierzcik, 1977; Yin, Heald, & Vogel, 1977), and "case surveys," which involve the quantification or structuring of observational data collected across many organizations (Royster, Baltzell, & Simmons, 1979; Yin *et al.*, 1977).

In this study, we have gathered both types of data (quantitative/structured data and qualitative/unstructured data) from both types of informants in order to capitalize on their respective strengths. If we combine the two dimensions—informant locus and type of data—we can derive a fourfold table showing our different data sources, which fall into each of the four cells (Table 2). The descriptions of our major data sources which follow are keyed to this classification.

Structured Data from System Members

The major sources of structured, quantificable data from system members were the Professional Personnel Census Forms. Starting in the fall of 1973, all professional employees of the 10 participating ES districts annually received a questionnaire that tapped attitudes about change, perception of the ES program, important features of their work environment (i.e., the organization's structure and culture), and their own background characteristics.[4] The questionnaires varied somewhat from year to year: many questions were repeated over time, some were deleted, and some items were added when deemed appropriate. The questionnaires were mailed to each professional staff member that was employed by the school district in a given year, rather than being distributed by the school administration, and were returned by mail to Cambridge. The analyses presented in this volume draw most heavily upon two of these questionnaire administrations: that which was administered in the fall of 1973, and the last instrument, administered in the spring of 1977. All batteries of items for the 1973 questionnaire that had produced reliable data and were appropriate for our analysis of system change were included in the 1977 questionnaire, in addition to many new items measuring the professional personnel's final assessment of the value and impact of the ES program.

The overall response rate for the 1973 questionnaire was 72% (N = 678). In 1977 it was 54% (N = 562). Several procedures were used to enhance response, including reminder letters, the distribution of a summary report to all staff members in the 10 districts in 1976, and a second mailing of the 1977 survey. Despite these efforts the response rate decreased; we attribute this fact to a variety of causes, including respondent fatigue, as well as high turnover in many of the districts so that in later years many staff members were relatively unfamiliar with the ES program.

Structured Data from System Observers: On-Site Researcher Questionnaires

The OSRs, who lived in each of the 10 districts from the spring or summer of 1973 through the summer of 1976, were important informants about several aspects of the system and filled out questionnaires

[4]A pilot questionnaire was sent to personnel in six districts in the spring of 1973. Except for the fall of 1973 (which was an anonymous questionnaire), all subsequent surveys were made in the spring of the school year and were *not* anonymous. Background characteristics were obtained only once, via the first spring questionnaire to which a staff member responded.

and checklists that were designed by the Cambridge-based project staff. In particular, the OSRs provided our study with two important types of structured data. The first was on system characteristics, such as types and number of employees in the school systems (job classifications), and some basic characteristics of the schools.

A second and more important type of structured data provided by the OSRs was in response to questionnaires that dealt with the degree of implementation of each district's ES project components, the degree to which each of the district's schools had actually implemented ES activities, and the degree to which the implementation of ES had produced the anticipated changes. We believed that the researchers, as observers of the system, would provide data on implementation and change that were considerably more reliable than that which would be reported by members of the system, since these "outside observers" had little vested interest in the success or failure of the ES program. Furthermore, our measures of implementation were based on some highly abstract concepts which we felt could be responded to best by trained social scientists accustomed to dealing with such abstractions. In addition, the forms for measuring ES implementation, particularly the forms that examined implementation for each component, were quite burdensome, and would have taxed the cooperative spirit of even the most research-oriented school administrator or teachers. In many cases, for example, the locally based researchers were required to interview several people in order to obtain the appropriate information to fill out a single form, since the data that we requested were not of the type that would normally be collected by the school district.

During the period when the on-site staff were completing the structured instruments, there was frequent telephone contact between them and the Cambridge staff, during which questions and interpretations were discussed. In addition, when the completed forms were received, they were read through by the senior author, and if there were any questions about responses the on-site staff members were contacted for clarification or elaboration.

OSR Questionnaires formed the major data source for the quantitative measure of implementation and change, both at the district and at the school level. (The issues involved in the conceptualization and measurement of implementation in the ES program are discussed in greater detail later in this chapter.)

Unstructured Data from System Members: Telephone Interviews

While the OSRs were in residence, it was a policy of the research team not to jeopardize their locally based rapport by engaging in direct interviewing of site personnel by Cambridge-based staff members. However, after the departure of the OSRs from the sites in the summer of 1976, some information was still needed to complete the study. In addition, the Cambridge-based staff felt a need to have direct contact with the site personnel, in order to understand firsthand the frames of reference that were used by the local school personnel. Thus, unstructured telephone interviews were conducted with superintendents and school principals in the 10 districts in the spring of 1978. These interviews elicited the respondent's assessments of the overall benefits and problems that had accompanied their district's participation in the ES program. Information was sought on the plans being made at the district and school level for the continuation of components or programs that had begun under ES. The telephone interviews were based in part on a very brief survey mailed to each respondent that asked which of the components implemented in their school or district were continuing, and in what ways continuation was being supported. Except for using the survey infomation as a means for probing continuation plans, the interviews were conducted without a formal-interview protocol; rather, a set of very general information goals were used to structure the conversation. The interviews lasted from 30 minutes to over 1½ hours depending on our research needs and the informant's desire to continue the conversation.

Training and a common understanding of terms and interview objectives between interviewers are particularly critical when conducting unstructured interviews of this type. As a consequence, preliminary interviews were done only by two staff members who had been involved with the project for several years. During their first few contacts with respondents, these two key interviewers worked together in order to develop a clearer understanding of the types of information that could or could not be easily obtained from the telephone contact. They then developed training instructions and subsequently trained four additional interviewers to conduct the interviews. Each of the four interviewers was debriefed after completing interviews for a given district. During the debriefing a quality-control assessment was made to evaluate each respondent and interview. Judgments about interview quality were used in weighing the usefulness of each interview for the analysis.

Unstructured Data from System Observers: The Case Studies

As we have discussed above, each of the 10 districts was the subject of an extensive case study written by a professional anthropologist or sociologist on the basis of three years' residence in the district. When we began to analyze our data, it became apparent that there were two alternative approaches to using these extensive documents as secondary sources. The first would involve developing a classification scheme for the variables and/or issues that were of concern to our study, and developing a coding, indexing, and retrieval system for locating those data as appropriate. A second approach would be to delegate the responsibility for immersion in case-study data to a single staff member, who could then serve as a resource for ensuring that qualitative information informed both the analysis and the writing at each stage. An important feature of these case studies, for our purposes, was that our study had no direct influence over the sources of data, the methods or techniques used to collect data, or the analysis and interpretation of data. As a consequence, although the case studies were a very rich source of secondary data for the study of organizational change in the ES districts and schools, in most cases only a few of the case studies would address a particular point that we were emphasizing in our research. (Some data were available from all or most of the case studies.)

Since none of the case studies approached the site from an explicit organizational change perspective, nor were they written to a common outline, using the case studies for our own analysis required that we be familiar with the work as a whole, so that we could turn to it quickly when it seemed appropriate to use the case studies to elaborate or expand upon the data sources that we ourselves had designed.

Thus, at various stages in the analysis, staff meetings would be held in which all persons involved in the analysis would discuss and interpret patterns of findings based both on case-study data and data from other sources. These interpretive seminars resulted in the greatest use of case studies in the analysis. However, in addition, when survey-based sections of the analysis were completed, the staff case-study specialist would read them and systematically attempt to apply case-study materials to the findings (either in support or to point up discrepancies).

In addition to the 10 case studies, we also made use of the data provided by OSRs for five of the sites in another published report from the larger project (Herriott & Gross, 1979) and from a special study of the ES/Washington program officers (Corwin, 1977). These reports have formed the basis for a large part of the analysis presented in Chapters 4 and 5.

STRATEGIES FOR MEASUREMENT AND ANALYSIS

As pointed out in Chapter 1, ES was not a true experiment with a clearly definable "treatment"; hence our approach to analysis has not been to look for treatment effects. Rather, we have viewed the emergent ES process as a field experiment in which the causal variables of interest are the process of planning and implementation of the local program on the one hand, and the preexisting system characteristics on the other. Our basic strategy has been to examine the relationship between, on the one hand, the nature of the districts and schools at the earliest possible point for which we have good organizational measures (in most cases the fall of 1973), and, on the other hand, the outcomes at a later point (a date which varies from the spring of 1975 to the fall of 1977, depending on the outcome measure being examined).

Our analysis strategy has always focused on organized units rather than individual occupants of organizational roles. Individual respondents, whether they were teachers or administrators responding to an annual survey or administrators responding to a telephone interview, have been treated as informants about the organization rather than about themselves as individuals. In all cases where there have been multiple respondents from a school or district (for example, the Professional Personnel Census Forms), responses from individuals have been aggregated to the organizational level in order to get an organizational measure of the variables in question.

Thus, for example, we have used teacher reports about the degree of influence of various role actors in the system to calculate a school score on, say, principal influence. This aggregation was done by assuming that the *actual* level of principal influence in a school could be adequately reflected by the average rating of principal influence given by all of the teacher respondents that were based in a particular school. In some cases, organizational measures have been more direct—such as size and complexity—and have not been based on data aggregated from individuals.

This study may thus be viewed as a panel study of *organizations* in which we have gathered many of the data from samples of individuals. Different individuals may have responded to the surveys at different times, because of staff turnover and response rates of less than 100%; what remains constant are the *organizational units* that are being studied.

Two different organizational levels interested us: the school and the school district. Since ES was designed as a district-wide program, part of our analysis was conducted at the district level. However, ES implementation occurred primarily in schools. This fact, coupled with our assumptions about the loose linkage of schools within districts, was found to be a compelling reason for pursuing much of our analysis at the school level. Analytic techniques did, of course, vary when the unit of analysis was the district or the school, because of the more limited number of cases in district-level analysis. (N of school districts = 10; N of schools ranged from 45 to 49 depending on data availability, as noted in Chapters 6 and 7.)

The four types of data-collection efforts described above (Table 2) yielded a large and complicated data base. The measures of system characteristics were derived in large part from the available literature on organizational sociology and the sociology of education (see Chapter 2). These measures—specifically, their indicators and how they were scaled—are discussed in greater detail in the relevant analytic chapters that follow.

However, one important measurement issue faced us for which there was little guidance in the available literature—measurement of the implementation of complex programs of planned change. Since implementation became a basic focus of our research, we found it necessary to elaborate the concept beyond the level found in the literature at the time this study was begun. Given the importance of this concept to our research, and the extensive use of its measures in Chapters 6–10, we have singled out for discussion in this chapter our approach to the measurement of implementation. In order to systematically measure implementation across 10 fairly diverse ES projects, we developed a new construct—a multidimentional "scope of implementation." This construct enabled us to arrive at more sophisticated conclusions than "Yes, implementation occurred," or "No, it did not." The concept of scope permits us to ask, "What areas of the educational systems were being altered? How much change was being implemented? What kinds of change were occurring?"

We conceive of the implementation of change as involving two vectors. One vector, which we have called "facts of educational change," refers to the *aspects of the educational system in which the change is taking place.* For this vector we have used the facets of comprehensiveness as outlined by the ES/Washington staff.[5] These facets are: curriculum; staffing; time, space, and facilities; community participation; and administration and organization.

The second vector of change concerns the *nature of the implementation that is taking place.* This vector comprises two dimensions of organizational change: the quantity of change and the quality of change.[6] Quantity of change refers to the degree to which the innovative efforts have penetrated the system, and its two subdimensions—pervasiveness and extent—identify different types of penetration. Pervasiveness represents the percentage of schools, teachers, pupils (or some other unit) affected by the change. For example, a change which only affects third-grade teachers is less pervasive than one affecting all elementary-school teachers. Extent refers to the percentage of an affected unit's total available time which is being altered. Thus, regardless of what proportion of the teachers are affected by a new program or other change, if only 50 minutes of the teacher's day are being affected, the change is less extensive than one which has an impact on the entire workday.

Quality of change, on the other hand, addresses the issue of how radical or different the change is compared to that which was replaced. Unlike the measures of the quantity of change, quality measures depend on a subjective judgment of whether the innovations actually result in alterations in content, behavior, or structures associated with

[5]The five facets of comprehensive change were deliberately designed to be specific to the ES program. However, such a vector of implementation that locates the change within a system is clearly modifiable to the study of implementation in other programs of planned change.

[6]Although the facets are specific to the ES program and this vector is modifiable for other programs, the dimensions are relevant to any program that attempts to make more than minor adaptations within a social organization.

Table 3. Scope-of-Implementation Matrix

Facets of educational change	Dimensions of educational change			Overall facet scores
	Quantity of change		Quality of change	
	Pervasiveness	Extent		
Curriculum				
Staffing				
Time, space, facilities				
Community participation				
Administration				
Overall dimension scores				Total scope score

the educational system. For example, a new French program that simply involves a change of textbooks may be only slightly different from the previous language programs available in the school, whereas one which involves introducing French language instruction at earlier grade levels, and which introduces language laboratory facilities, would represent a more dramatic change.

In order to begin to measure comprehensive organizational change, we constructed a "scope" matrix, displayed in Table 3, in which facets intersect with dimensions to produce 10 separate facet/dimension cells (15 when the subdimensions of pervasiveness and extent are included), each representing a distinct operational measurement of change. Thus, for each of the five facets, we assess the scope of implementation in terms of the two dimensions of change. In addition, five marginal cells (the right-hand column) represent overall change in each of the five facets, and two marginal cells (the bottom row) represent change in each of the two dimensions. On the lower right, a cell representing a grand, overall measurement of the scope of change summarizes all the cells in the matrix. The specific methods we used to operationalize this matrix in order to measure scope of implementation at the school and district levels will be described in Chapters 6 and 8.

Chapters 1–3 have set the stage for understanding a complex process of planned change, the Rural ES program. Chapter 4 introduces the protagonists (and sometimes antagonists) in this process—ES/Washington and the 10 local school districts—and begins the story of the change effort.

4

On the Brink of Change

The Federal Level and the Local Level

One of the major linkages that affected the course of the ES process of change was the relationship between the federal sponsoring agency (The National Institute of Education during all but the very early months of the program) and the local, rural school districts. The variety of targeted federal programs intended to stimulate change and improve the quality of local services has multiplied in the last two decades, particularly in the educational sphere (Bailey & Mosher, 1968; Reagen, 1972). Examining the federal-local relationship which developed during the life of the Rural ES program can help us understand the impact of this growing trend on local programmatic efforts.

Although the increase in federal support for education has been welcomed by many as an opportunity to equalize resources and expertise, it has also raised concerns about issues of power, autonomy, and locus of control. As federal dollars for education have increased, so has federal influence on the local educational process, resulting in a struggle to maintain a working balance between national and local interest. This struggle has been an issue for federal programs of all types, but it has been particularly sensitive in education, given the American tradition that authority for education lies with the states and their agents—the local school boards (Berke & Kirst, 1972; Corwin, 1977).

However, the federal-local linkage within educational programs is not only a function of the struggle for power and authority; it is characterized by normative considerations as well—that is, by the degree to which the definitions of the program, its assumptions, and its goals are shared by the participating actors at both levels. Typically in federal intervention programs, the assumptions of the federal program planners are poorly explicated, and there may be a lack of mutual understanding of what is to be the federal-local relationship. As a result the expectations of the various "partners" may diverge greatly, and the consequences may be not only a clash of authority but also a failure to achieve the desired outcomes of either partner (Pincus, 1974).

The ES program can be viewed as a partnership in educational change, where the federal program office (and its program officers, who were the links with the local school-district personnel) constituted a distant but very salient environment for the local districts. This chapter introduces the partners in this process, first by considering the federal program and the assumptions on which it was based and the implications of those assumptions, then by describing the 10 participating school districts and their apparent readiness for change.

ASSUMPTIONS UNDERLYING THE EXPERIMENTAL SCHOOLS PROGRAM

In order to understand the federal-local relationship that was intrinsic to the ES program, it is important to examine the educational and political context from which the program emerged and to identify the assumptions on which the program was based; these had a lasting impact on the relationship and the entire innovative effort throughout the life of the program. Every effort at planned change—ES is no exception—develops on the basis of a set of assumptions. Some of these are likely to be explicit, conscious, and clearly intended, and to be founded on documentary evidence.[1] However, a program may also contain within it some implicit assumptions, less clearly articulated,

[1]For the ES program the main basis by which we identified explicit assumptions consisted of the Announcement, Congressional hearings for budget justification, and a memorandum written by one of the designers of the program (Budding, 1972). We have also drawn from the findings reported by Doyle, Crist-Whitzel, Donicht, Eixenberger, Everhart, McGeever, Pierce, and Toepper (1976), which included an extensive investigation of the early documents in the ES/Washington files.

perhaps "talked about," but not documented. Some of these may be known only when they are logically deduced from circumstantial evidence of their existence. Sometimes they are written down *ex post facto,* either by the program designers themselves or by students of the program attempting to gain a clear understanding of the phenomenon and its consequences.

Explicit assumptions are generally expected to lead to certain anticipated outcomes, while the effects of implicit assumptions cannot necessarily be predicted although these effects are often just as important. For the ES program, both types of assumptions appear to have contributed to the process of planned change and its outcomes.

For the ES program, as well as other federally sponsored programs, three general categories of program assumptions can be identified: assumptions about the nature of the process that the innovation goes through (in this case, the process of planned change); assumptions about the nature of the intervention or treatment (in this case, the design of the ES program); and assumptions about the target of the innovation (in this case, the local rural school districts). In the discussion which follows, we briefly describe the three general categories of assumptions (both explicit and implicit) about the ES program.[2] We also discuss the effects that some of the assumptions had on shaping the federal-local relationship, particularly during the planning year.

Assumptions Concerning the Process of Change

One type of assumptions underlying the ES Program concerns the nature of the process of planned change and the role that the federal government could play in that process. Five such assumptions can be identified.

1. *Previous change efforts had been a failure, and their failure could be attributed in large measure to their piecemeal nature.* The many change efforts sponsored by the federal government which began in the 1950s and proliferated in the 1960s were found to have effects and impacts that were minimal or nonexistent. It was the perception of the ES designers that the federally funded change efforts of the past had

[2]Much of this discussion is derived from Herriott and Gross (1979), particularly chapters authored by Gideonse, Kirst, Lippitt, and Herriott. We particularly acknowledge the contribution of Gideonse, who first articulated the notion of explicit and tacit program assumptions and of several of the assumptions discussed in this chapter.

lacked overall coherence, and had fostered a piecemeal change strategy as well as an emphasis upon the development of new educational products (curricula, techniques, and machines) inappropriate for widespread utilization at the local level (Budding, 1972).

The judgment that the cause of the failure of previous change efforts was their piecemeal nature was perhaps the most important assumption made by the ES planners, for it led to the decision on their part to build the ES program around the overall theme of comprehensiveness. The design of the program thus seems to have been, at least in part, a reaction to the perceived failures of the past. If those programs had been piecemeal, this one would be comprehensive. The elements of comprehensiveness were defined in an "Announcement of a Competition for Small Schools Serving Rural Areas," which was sent by the Department of Health, Education and Welfare to all school districts in the United States having fewer than 2,500 students:[3]

> A small rural Experimental School's project must be comprehensive, that is it must include at least the following components:
>
> (a) a fresh approach to the nature and substance of the total curriculum in light of local needs and goals;
> (b) reorganization and training of staff to meet particular project goals;
> (c) innovative use of time, space, and facilities;
> (d) active community involvement in developing, operating, and evaluating the proposed project; and
> (e) an administrative and organizational structure which supports the project and which takes into account local strengths and needs. (The Announcement, p. 2)

These key elements became known as the five "facets" of comprehensiveness. The requirement of comprehensiveness did not necessitate the total replacement of everything being done with something new, but it did mean that ". . . what is going on in each of these areas should be related to, consistent with and supportive of all the other areas" (The Announcement, p. 2).

2. *Comprehensive change would have a total effect greater than the sum of its parts.* The ES designers assumed that combining innovations into comprehensive packages would produce multiplicative results. The National Institute of Education called this multiplicative effect "synergistic" and considered it to be one of the major concepts which the ES

[3]Throughout this and subsequent chapters, this important document will be referred to as "The Announcement." It is reproduced in its entirety in Herriott and Gross (1979).

program was designed to test (Doyle *et al.*, 1976, p. 31). Schooling was viewed as a total program, and the attainment of educational changes required change activities in all aspects of the educational system. Although the ES project was to be operationalized through a set of project "components," it is unclear whether there was actual consensus among the ES designers as to whether comprehensive change was an intended *outcome,* thereby implying the transformation of a school district, or a *strategy* for change—one that tries to force enough simultaneous change activity to overcome the system's inertia.

3. *The change process is a rational one that can be managed effectively by school personnel.* The ES designers assumed explicitly that change is a rational process, and that, given clearly articulated goals and a broad-based planning phase (in the case of the Rural ES program, the planning process was intended to continue for almost a full year), rational choices would be made, and implementation and routinization of change would be management tasks.

However, as has been suggested in Chapter 2, the rational-change process is generally moderated by nonrational elements. The looseness of both structural and normative linkages can militate against a clearly ordered management system, and the achievement of stated outcomes is heavily dependent upon the consensual commitments of the various role partners in the process.

4. *A federal stimulus is necessary to effect widespread change at the local level.* The entry of the federal government as a stimulus and participant in the process of local educational change appears to be based on the assumption that local schools are in need of improvement but that a combination of factors appear to be preventing *self-initiated* change efforts. These factors include the inability of local school districts to muster the slack resources necessary to undertake innovative efforts, the resistance of taxpayers to increased local expenditures on schools, and the lack of indigenous expertise required to facilitiate change. Thus, some outside stimulus for change is often necessary (Griffiths, 1967), and this role can appropriately be assumed by agencies of the federal (and state) governments.

5. *The federal government either knew how to facilitate change at the local level or could learn in a timely fashion.* ES/Washington[4] allowed the

[4]In the remainder of this volume we refer to the federal-level administration of the ES program as "ES/Washington."

local districts a full year of extensive planning before beginning implementation of their projects; however, it placed no such requirement upon itself. The implication is that ES/Washington assumed that it possessed the capacity to carry out the requisite duties of the program (Gideonse, 1979). This assumption can be viewed as one which contributed much to the lack of clarity which characterized federal-local communication.

Evidence from Corwin's (1977) study of program officers involved in the ES program (who were primarily responsible for carrying out ES/Washington's responsibilities in the field) supports the conclusion that there was inadequate planning at the federal level for its facilitative role. However, there were many circumstances in the Washington bureaucracy that acted to prevent such planning from occurring even when the original program sponsors wanted or intended it to. (See Sproull, Weiner, & Wolf, 1978) for an extensive examination of the National Institute of Education's early years.)

Even while the competition was being designed, plans were being made for the transfer of the program from the U.S. Office of Education (OE) to the newly authorized National Institute of Education (NIE). This move was accomplished in August 1972 and had several implications for the program. To NIE, ES was a stepchild inherited from OE, which NIE had not designed but for which they had become responsible. The transfer also meant a loss of support over the program by its original sponsors, many of whom remained at OE. Therefore, even though the original program sponsors envisioned building a capacity at the federal level for facilitating local change, such plans were inadequate for surviving the transfer of responsibility from one federal agency to another.

Assumptions about the Program Design

A second set of assumptions relate more specifically to the program objectives than to the process of change per se. These assumptions were prominent determinants of the federal-local relationship and had a great deal of impact on the local planning process. Five assumptions of this type can be identified.

1. *The program would have a dual focus: research and grant-in-aid.* One of the stated purposes of the ES program was to discover ways to improve rural education:

> The Experimental School program, through this competition, is making available the opportunity for a limited number of rural school systems to test new ideas for educational improvement which are developed in and for a small, rural school setting. Since the number of projects that can be funded is limited, and since what can be learned from them may be valuable for many small school districts, it is important to have these efforts thoroughly documented and evaluated. (The Announcement, p. 1)

One way in which the research focus was manifested was the awarding of a separate contract for external evaluation and documentation (i.e., the contract under which our research was conducted). The ES program also required each district to build into its project local methods for documentation and evaluation.

Underlying the explicit assumption that the program would have a dual focus (research and grant-in-aid) was an implicit assumption that the two objectives would not interfere with each other. There is some evidence, however, that this implicit assumption caused problems in the program's functioning. Much of the burden of resolving the inherent confusion resulting from the dual focus of the program fell upon the federal program officers who had the responsibility for monitoring the local projects. On the one hand, the grant-in-aid nature of the ES program implied to them a federal commitment to facilitate the success of the local projects by providing them with extensive technical assistance. On the other hand, the research focus of the program implied to many program officers that the proper posture was "hands off" the local projects, even if the projects appeared to be failing (Corwin, 1977). This ambiguity in the nature of the monitoring process characterized the federal-local relationship through most of the project years.

2. *Comprehensiveness would become defined as the program evolved.* While the requirement of a comprehensive project was explicit, it seems that federal program planners held the implicit expectation that the meaning of comprehensiveness would become clear as the program evolved. The outcome of mutually supportive innovations in all facets of all schools in a district was expected to be transformation of the district. This transformation was to take place, however, without disruption to the normal processes of schooling. At an early stage in the ES program, an ES/Washington program officer likened the process of comprehensive change to a home-improvement process during which the entire house undergoes renovation. Such a process does not involve the construction of a totally new "house," nor does it require

demolition and rebuilding. Instead, it takes an existing structure (a school district) and, while still "living in it," conducts massive renovation.

3. *The entire school district was the appropriate unit of change.* In order to achieve the goal of transforming the district, the rural ES designers considered the entire school district as the unit of change. Although it was not explicitly stated in the Announcement that school-by-school planning was unacceptable, differential planning at the school level was discouraged. Instead, despite the commitment of ES/Washington to sponsoring and testing "different" programs within the ES framework, the different programs were all to be planned at the district level. Kirst (1979) refers to this assumption as a strategic miscalculation which flies in the face of recent literature on implementation, stressing the school as the optimal unit of change.

Furthermore, this assumption implied the ability of school-district administrators to manage a coordinated, district-wide program in schools that may, in fact, be loosely linked due to demarcations of different school levels and to the relative autonomy of schools regardless of level. In rural school districts, the physical distance between schools serves to loosen the linkage among them even further. It became highly problematical for the managers of the change process to overcome the effects of loose linkage when trying to treat the district as the unit of change. Although planning was conducted at the district level, as implementation progressed the school clearly became the unit of implementation. (Implementation at the school level is the focus of Chapters 6 and 7.)

4. *External funds (i.e., ES dollars) could be used to develop and implement a project that could continue with regular school-district resources.* Many federally sponsored programs for change are in the form of "seed money," that is, small amounts of money to help stimulate changes. Other federally sponsored programs—such as Title I—have been in the form of "support money," in which the monies are used as continuous support for special services. A third type of program has been in the form of "program money"; in this case, funds are allocated for a specified time to help in the solution of particular problems. (Many of the Emergency School Aid Act funds have been used in this way.)

The ES program can be classified as program money of a special type. A very explicit intention of the ES designers, and one that was frequently reiterated during the planning year, was that the districts must assume financial responsibility for maintaining the innovations once the funding period was over. Thus ES funds were considered to a "bubble" in the budget, one that would provide special resources to develop and implement innovations that would not require extraordinary funding to sustain. The implicit expectation was that the funds would indeed be used in this "facilitative" way and not to support special programs or ongoing services. As the projects progressed, this provision was either not clearly understood or not adequately monitored by ES/Washington, for, as will be described in Chapter 8, each of the 10 school districts used at least some of the funds in other ways and was not able to continue all the innovations after the funding period ended. This may have been due, in part, to the "hidden agendas" in each school district; many of the districts saw ES as an opportunity to acquire funds for special programs they could not afford with existing resources. Further, in order to gain support of teachers and parents for the project, it was often important to use some of the funds to implement popular or highly desired special programs.

5. *Criteria such as "comprehensive change" and "locally defined plans" are complementary objectives.* Two of the main objectives of the ES program were that education would be improved through comprehensive change and that each district's project would be designed locally. Although these two objectives were not considered to be contradictory, they, in fact, turned out to be somewhat conflicting. The ES designers and program officers envisioned during the planning year a concept of comprehensiveness which they felt should be reflected in each school district's plan. On the other hand, local school-district administrators either failed to understand the concept, chose to ignore the extent to which they would be expected to disturb their educational systems, or responded to the concept of local planning with the understanding that it was an opportunity to get involved in an effort "with no strings attached." The notion of local planning along with a five-year advanced-funding commitment seemed to imply freedom from federal control. However, the school districts found during the planning year that if their locally designed plan did not meet the federally defined criteria of comprehensiveness, as interpreted by the program officers, they were required to revise their plans to meet those re-

quirements. This lack of normative linkage (mutual understanding) of the program objectives created many problems for the districts very early in the five-year process.

Assumptions about the Local School Districts

A third set of assumptions which formed the basis of the Rural ES program concerned the nature of the rural school districts themselves and their ability to engage in an enterprise as complex as the one envisioned by the ES designers. Four assumptions can be identified.

1. *The local school districts had or could independently acquire the expertise to carry out their projects.* A major theme of ES from its conception was that of locally devised solutions to locally perceived problems. The ES program was not to be another instance of a federal agency's persuading school districts to accept a federally endorsed innovation to solve local problems. This expectation rested on the implicit

> assumption that a local educational agency *could* design and operate a program which would alleviate a self-defined problem through extensive alterations in existing methods of operation, without undue intervention by the funding agency. (Doyle *et al.*, 1976, p. 50) (emphasis added)

As a consequence, the program did not include the explicit provision of technical assistance to the local districts for planning and implementing their projects such as that which is often provided by other federally sponsored programs for local problem solving—for example, the National Diffusion Network or the Research and Development Utilization Program.[5] Instead, the districts were expected to use ES funds, if necessary, to acquire such assistance. Some of the program officers periodically performed a technical-assistance role in addition to their contract-manager or advisory role (Corwin, 1977), although the intensity of such involvement was limited, in part because of the conflict of the dual focus of the program—research and grant-in-aid. The provision of assistance, it was feared, would "contaminate" the research.

In reality, however, many of the districts lacked the expertise either to carry out their projects with indigenous staff or to acquire the

[5]The original Request for Proposal and the contract between ES/Washington and Abt Associates for the external evaluation did include the specification of evaluative technical assistance to the sites. However, a contract modification during the first year of the research eliminated this provision because of its inherent conflict with the broader research objectives of the study.

necessary assistance independently, despite the availability of ES funds for such activity. As a result, for many districts, ES turned out to be an underutilized opportunity to achieve the desired solutions to local problems.

2. *The local districts understood the planning process well enough to make constructive use of the planning year.* One element of the ES which was believed to be particularly valuable in achieving the goals of the program was a planning year supported by a federal grant to each district. The concept of a year devoted to local planning is obviously related to the assumption that the districts had (or could quickly acquire) the expertise required for such planning and for building local support for their projects. This implicit assumption appears to have been ill-founded. Although ES/Washington provided substantial one-year planning grants (ranging from $45,500 to $120,400), the local school districts experienced serious difficulties in designing viable plans, which led one observer to characterize the planning year as "one of false starts and unrealistic expectations" (Kent, 1979, p. 327).

3. *Local community involvement would not intrude on local school-system autonomy.* The ES designers hoped to encourage participation and commitment of the community and give to the local communities as well as the school-district personnel a sense of ownership of the projects. It appears that ES/Washington envisioned community involvement and local autonomy as complementary objectives. As documented in individual site cases studies (Herriott & Gross, 1979, Chapters 4–8), they were never able to operate as such and, in fact, community participation was perceived by administrators in the local district as an intrusion on their own autonomy. Although ES planners in Washington saw local autonomy inclusively (i.e., to involve the total school-district population), local administration apparently saw it exclusively (within only the core planning group of professional educators).

4. *Rural school districts are similar to each other.* Although this implicit assumption may appear to be counter to the explicit ES program theme of local design and local autonomy, there is evidence that ES/Washington lacked sensitivity to the many differences among the 10 rural districts involved and held similar expectations for all of them. For example, in the original Announcement, it was specified that all participating districts must include a K(1)–12 student population. However, the structure of rural school districts does not always con-

tain all grades. A small rural district may contain only an elementary school and send its pupils to a neighboring district for secondary education. The Annoucement suggested that, in order to meet the K(1)–12 requirement, several rural school systems could join together in making an application—although such partnerships could be dissolved at the end of the ES project. This suggestion overlooked the strong sense of local control of many rural areas and the rivalries that often exist between neighboring districts. School-district consolidation, for example, has often met with community resistance (Tyack, 1974) as school-district boundaries lacking correspondence to natural community boundaries are created. Partnerships in two of the ES districts formed for the purpose of planning ES projects did not last through the planning year. In a third district, the attempt at collaboration presented many obstacles in designing and implementing a project.

ES/Washington also assumed that each of the districts could follow a similar timetable to conduct needs assessments, prepare viable project plans, plan and conduct internal evaluation, prepare management reports to NIE, and so forth, and that the requisite skills were either available or could be obtained in all the districts. However, this was not the case.

Areas in which the districts were dissimilar in ways not taken sufficiently into consideration by ES/Washington included their level of sophistication (particularly in dealing with agencies of the federal government), differences in population density (i.e., their isolation), differences in their history (e.g., recency of consolidation), differences in local values, and differences in the economic conditions and patterns of growth or decline.

The assumptions described above, then, form the background of the federal ES program. They are the basis for understanding the interactions which took place as the federal-local relationship unfolded, and for clarifying the many misunderstandings and disappointments which subsequently arose.

THE TEN RURAL SCHOOL DISTRICTS AND THEIR READINESS FOR CHANGE

How did 10 rural school districts become involved in the ES program and how ready were they to embark on such an ambitious endeavor? Past research on the process of planned change suggests that there must be a degree of readiness to change on the part of both individuals and organizations if the process is to proceed successfully (Glaser & Ross, 1971; F. Mann & Neff, 1961; Rogers, 1962). Ideally such readiness is thought to involve the existence of a perceived need to change, the availability of external resources to facilitate change, and the willingness to apply these resources to meet such needs. In reality, however, the impetus for change may operate in the absence of an awareness of particular needs within a school district. After studying 29 projects of planned change, Greenwood *et al.* (1975) concluded:

> Most schools store their needs in a bottomless pit. When they become aware of resources which could be matched with *any* given need, they fish around in the pit, find the need and tell the funding agency that that one was the most pressing thing on their agenda. (p. 27)

Regardless of whether the availability of resources leads to the definition of a need, or the existence of a particular need leads to the recruitment of external resources, a readiness to change seems to require some linking of needs and resources. It also seems to require a willingness on the part of the organization to enter into a relationship with an external agency whose objectives in facilitating change may be unclear until the process is well under way. School districts seem to differ greatly in their ability to systematically link needs and available resources and to understand the agendas of external agencies. However, the degree to which this has been accomplished seems to be an important indication of the readiness of school districts to embark upon the change process. It is our assumption that such readiness can have important implications for effective accomplishment of the subsequent stages of initiation, implementation, and continuation.

The Entry of Ten School Districts

The formal starting point for the participation of the 10 school districts in the ES program (and subsequently in our research) was their initial awareness of the program. In general, this awareness was achieved through the receipt of the Announcement in March, 1972.

Those school districts interested in planning and implementing a project of this type were requested to submit a "letter of interest" of not more than 15 pages to the U.S. Office of Education by April 15, 1972—four weeks after the receipt of the announcement. Approximately 320 of the more than 7,000 eligible rural school districts submitted such a response.

In asking applicant school districts to prepare a letter of interest, the ES program was seeking a rational basis for judging their readiness to undertake a process of planned change. A series of topics were to be discussed in the letter. Of particular interest to the federal agency were the responses of the applicant districts to the following instructions:

> Describe your current educational program stating present educational purpose and goals, the nature of the curriculum and the present organization to accomplish stated purpose and goals.
>
> List and describe what you feel to be the most important strengths and resources available to your school district which would be most helpful in developing an improved educational program. Please rank them in order of importance.
>
> List and describe briefly what you feel to be the most significant weaknesses of your current educational program, ranking them in importance.
>
> Describe how you would change and improve your educational program, utilizing the strengths and resources you have identified in attempting to overcome your weaknesses. (The Announcement, p. 7)

The letters of interest were subjected to a complex, multistage review which included the participation of ES/Washington staff members, regionally based panels, consultants, and site-visit teams. Each letter was initially read and rated by at least two ES/Washington staff members. Five external, regionally based panels were then given a sample of letters for review. From the resulting ratings, 25 "finalists" were selected, which were then subjected to further review by a committee of 13 consultants. Based upon the consultant review, three-member teams made site visits to 13 school districts. Ten geographically dispersed districts were ultimately selected (Figure 1). Six districts were awarded one-year planning grants with the "moral commitment" of four additional years of funding. Six other districts were awarded planning grants with the understanding that continued funding would be conditioned by the results of the planning process, and two of these districts were dropped from the program at the end of the planning year.

The letters of interest prepared by each of the 10 selected districts, although hurriedly written because of the four-week deadline for submission, provide considerable insight into their readiness for change. Each letter was unique in style and flavor. Yet, as a whole, they sounded a common tone of frustration with the quality of the

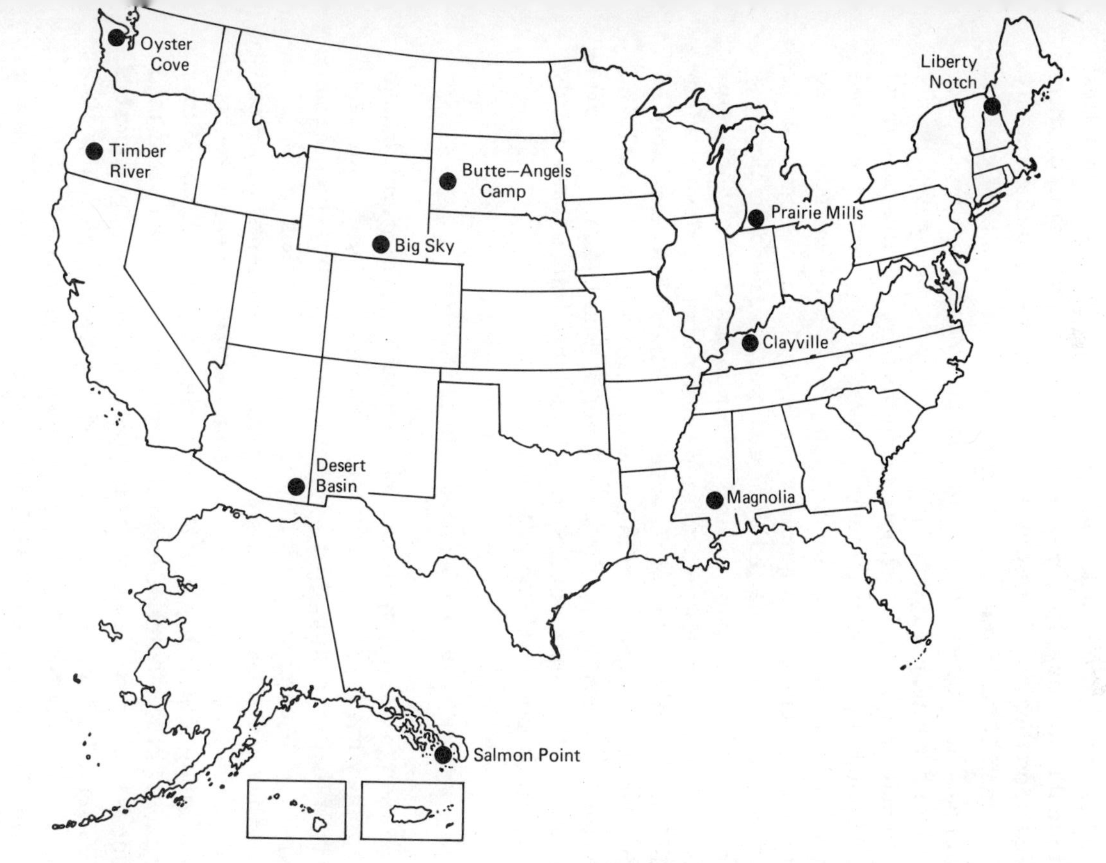

Figure 1. Geographical location of the 10 school districts participating in the Rural Experimental Schools program.

education these school districts were able to offer. Their problem, stated in broadest terms, was a series of "lacks," all of which were related to one another—lack of money, lack of trained personnel, lack of student motivation, lack of alternatives. Most of the school districts looked upon their human resources—that is, people in the community and in the administration, the teachers and pupils—as their greatest strengths. Their overall theme for change was clearly one of individualizing instruction and making it more appropriate to each pupil.

In many respects, the schools' strengths were also their weaknesses. Although the students, teachers, staff, and community were seen as resources, they were frequently identified as people needing to be motivated, retrained, and more involved. The rural character of the districts was as often referred to as a strength—providing closeness to nature, to other people, and an opportunity for wholesome activities—and often as a weakness, resulting in isolation, stagnation, and a low self-image. The very change to which the districts committed themselves by applying for funding was welcomed, anticipated, and aggressively looked for; at the same time, in their letters the districts emphasized the importance of their traditional values, close family relationships, and stable populations.

The Ten Selected School Districts

Although, as is often the case with federal program, it is difficult to determine what the exact selection criteria for the planning grant were, it is clear that the districts that were involved in the Rural ES program were, in all likelihood, atypical of rural school districts. In the first place, they were part of a very small group whose interest was stimulated by the original invitation to submit a letter of interest—a group representing less than 5% of all rural school districts. Second, they were able, in this letter of interest, to portray themselves and their needs as being particularly suited to a program with the characteristics of ES. Finally, they were able to survive what turned out to be a difficult planning year, which required them to produce a detailed plan for comprehensive change. While we have no way of making scientific comparisons with any other population of school districts, it seems reasonable to assume that those involved were probably higher than average on initiative and incentive to make alterations in existing programs.

On the other hand, there were also enormous variations between the districts along a number of dimensions relevant to the change process. Some districts were familiar with federal funding for program improvement, and employed administrators who were practiced "grantsmen," whereas for others this was their first real contract with externally funded innovation programs. Some districts were characterized by conservative and stable populations, who were deeply suspicious of the notion of comprehensive change, but others were undergoing significant population shifts, bringing a more affluent set of residents and new demands for improved schooling.

The availability of district-based administrative leadership to guide the ES program through its planning and implementations also varied. Districts endowed with sophisticated and influential chief administrators (one of whom left the district during the program to complete a doctoral degree at Harvard University) clearly differed from those that emphasized traditional patterns of school autonomy and parental control.

In sum, the 10 selected school districts represent a diverse group not only in terms of their geographic dispersion, but in their size, population density, economic base, history, problems, predominant values, and the characteristics of their staff. In this section, we present a brief description of each school district and its salient characteristics at the time it began its ES project.[6] (All district names are pseudonyms.)

SALMON POINT

Salmon Point is the smallest and the most isolated of the 10 districts. Located on an island in southeastern Alaska, it has no paved roads and is served by charter flights, daily amphibious flights, and a weekly boat service. Its population in 1970 was 300 people, of whom 50% were Indians. (However, during the project years, the population increased by 25%.) Employment is

[6]Population figures are for 1970, and number of students for the 1972–1973 school year. The main sources of information for these descriptions are the social and educational histories of each community (Fitzsimmons *et al.*, 1975) and our Professional Surveys of 1973 and 1974. In order to standardize cross-site comparisons of per-pupil expenditure for regional cost-of-living differences, the per-pupil-expenditure data reported by each district were adjusted using consumer-price-index (CPI) data for nonmetropolitan areas in the fall of 1973. See U.S. Department of Labor (1974), p. 59.

seasonal and uncertain but a growing lumber industry may bring with it economic and cultural changes.

The school population was 112 students, served by a staff of 19 in two school buildings. At the time Salmon Point applied for ES funds the annual school budget was $312,000, with a per-pupil expenditure of $1,215.

Teachers in Salmon Point had considerably less teaching experience (an average of 2.4 years) than teachers in the other districts. The district staff had a comparatively high morale, although Salmon Point experienced the highest staff turnover of any of the ES districts over the course of the project years. In comparison with the other ES districts, both the superintendents and teachers were seen as having a great deal of influence in decision making. The school board had a comparatively low amount of influence.

At the time of the Rural ES Announcement, the superintendent had already begun a program of change in the schools and in the letter of interest stated his interest in continuing it. In the letter, the superintendent also spoke of the isolation of the community's children, who grew up unaware of any culture but their own.

PRAIRIE MILLS

Prairie Mills is one of the more industrialized rural communities which participated in the ES program. Half the labor force in 1970 was engaged in manufacturing, although farming is still an important industry. Located in southwestern Michigan, the school district was consolidated nine years prior to its participation in the ES program and is located in parts of two counties containing one incorporated village and rural areas. Its population at the time of the 1970 census as 5,038, and the school district served 1,645 students in five schools with a staff of 92. In 1971–1972, when the district applied for ES funding, the total budget was $1.26 million, with a per-pupil expenditure of $787. The largest employer in the district was the school system.

Teachers in Prairie Mills had a significantly higher educational level than that of teachers in other districts, but were similar to teachers in other districts in age and number of years of teaching experience. Morale in this district was a serious problem, with almost all teachers noting that the inadequacy of appropriate instructional materials and supplies was a major problem. The most important educational goals identified by teachers were that students think for themselves and that students get an education geared to their individual needs. Principal authority was particularly strong.

Shortly before entering the ES program, Prairie Mills students had been administered a battery of tests as part of a state-wide testing program, which showed their overall self-esteem to be low. The theme of the letter of interest was built on this finding, and on the district's desire to develop a program which would pay particular attention to students who were not experiencing success.

BIG SKY

Big Sky has the largest geographical area and one of the most scattered populations of the ES rural districts. Consolidated from five school districts on the eve of its entry into the ES program, Big Sky comprises an area equal in size to the state of Connecticut; it is located in south central Wyoming and contains nine towns. Sheep and cattle ranching and mineral extraction are major industries and some tension exists between ranchers in the southern part of the district and miners in the north. Residents are united, however, in mistrust of the federal government and Easterners in general. The population in 1970 was slightly over 4,000, but was rapidly growing because of expansion in the district's coal-mining industries. There are 12 schools, and a staff of 129 were serving 1,294 pupils. At the time ES began, the annual school budget was $1,669,666 and the expenditure per pupil was $1,329.

In the letter of interest, which was written primarily by the assistant superintendent, the theme was one of using ES as a catalyst for unity: the program was seen as an opportunity for five previously autonomous school districts to work together to overcome the problems caused by the large geographic area and low population density.

CLAYVILLE

Clayville, which is located 90 miles from Louisville in northwest Kentucky, had recently experienced industrial expansion which caused a 32% growth in population from 1960 to 1970. Previously an agricultural area, the employment pattern became one divided between farming and manufacturing. The economy continued to expand slightly during the ES years.

The Clayville school system underwent consolidation 14 years prior to the ES project. The population in 1970 was just over 7,000. There were 1,490 students enrolled in four schools with a staff of 78 (a fifth school opened the second year of the project). At the time Clayville entered the ES program, the annual school budget was $1.243 million with a per-pupil expenditure of $739.

Teacher morale was particularly high in Clayville, which experienced the lowest turnover rate of any of the 10 districts during the five years of the project. Much of the stimulus for participation in the ES program came from the influx of new workers, who came from more sophisticaed communities to work in the new industries and who initially found Clayville schools inadequate. Many of these workers chose to live outside the county and were a tax loss to the county schools. It was the industrial consortium, whose interests lay in making Clayville an attractive place for their workers to live, that commissioned a private research organization to carry out an extensive needs assessment of the county school system. The school board subsequently took action by hiring a dynamic new superintendent from outside the district whom they perceived and who perceived himself as a "go-getter." He promised to build Clayville a school system they would be proud of.

The arrival of the Announcement of the ES program into this milieu was very timely and the superintendent responded eagerly. Not trusting the local expertise, he secured consultants from a nearby university to help write a letter of interest based on the findings on the recently completed needs assessment.

BUTTE–ANGELS CAMP

Butte–Angels Camp consists of two South Dakota communities in the west central edge of the state. The major source of employment is the local gold mine, and the local economy fluctuates with the price of gold. In 1970 the county had a population of just under 10,000, of whom 7,649 lived in the two towns of Butte and Angels Camp.

The school district was consolidated one year prior to the ES project, with a staff of 155 serving a student population of 2,217. The 1971–1972 school budget was $2,531,900, with a per-pupil expenditure of $1,176. The number of students served puts Butte–Angels Camp at the upper end of the ES districts in size, and there was some feeling on the part of ES/Washington that the district was really too big for the rural schools competition. (Butte–Angels Camp did, in fact, apply for the urban program of ES, before the Rural Announcement, and was rejected for being too small.)

The theme used by the superintendent in his letter of interest was based on an individualized-education technique called diagnostic-prescriptive instruction (d/p). The superintendent was considered by the local staff to be particularly "ambitious," and there had already been some tension between him and the district's teachers, who felt that his pursuit of "improvement" had brought them longer hours and more work without much reward.

LIBERTY NOTCH

Liberty Notch is not technically a school district but rather a governance unit (known as a Supervisory Union) of three separate school systems in northern New Hampshire. The superintendent who serves the Union is an administrator whose authority is otherwise extremely limited. At the time of its entry into the ES program, Liberty Notch was classified as an economically depressed area, with 23% of the families earning less than $3,000 per year. The primary source of employment is in forest and forest-related industries. Thirty percent of the population is either foreign-born or born of foreign parents, including a rather large French-Canadian population. The district population was approximately 3,500 in 1970.

A total professional staff of 91 in four schools serve a student population of approximately 1,150, some of whom come to the schools from neighboring Vermont. Teachers in Liberty Notch have a relatively low educational level, with very few having more than a Bachelor's degree.

The letter of interest was the work of one principal in one of the school districts involved in the Union, and was intended as an application for his own district only. His theme was one of developing a "model school" to serve as a community focal point. Students were to be involved in work–study vocational programs, and education would be made relevant to the learner and to the realities of the environment. A central part of the theme was the use of the school to serve the community by providing psychological services which were, according to the letter, desperately needed, because of the bleak environment and lack of opportunity for people in their surroundings.

In retrospect, it is evident that for Liberty Notch, the insistence of ES/Washington that the three school districts must become partners in the project and function (as they ordinarily did not) as a single unit was a problem from the outset. While there was some reluctance to form the partership, the promise of so much money brought the other two districts of the Union into the program.

MAGNOLIA

Magnolia school district is located in the southeastern corner of Mississippi. The county is about 80% forested, and the primary sources of employment are forestry and farming. In 1970 the population was 9,065, approximately 74% white and 26% black.

The school system was integrated without court order in 1970. At the time of the ES competition the district served 1,555 students (of whom one-third were black) in six school buildings with a staff of 119. The annual school budget was $758,000, and the per-pupil expenditure was $672.

A major inpetus for Magnolia to enter the ES competition was its recent desegregation, which made visible the disparity of the formerly separate systems. Critical reading deficiencies were found at every grade level, and the poverty of the district made it difficult to obtain appropriate instructional materials. The letter of interest was based on a recently completed, formal needs assessment which had been conducted by school district personnel with assistance from a university consultant. The letter acknowledged that the school system was basically disorganized and in need of change in staff, equipment, and teaching materials. The letter of interest was composed almost single-handedly by the curriculum director, who had the expertise to compose the document but who left the school district shortly afterward.

OYSTER COVE

The two communities which comprise the Oyster Cover district are located in the northwest part of the state of Washington. Logging and related industries are the major source of employment for the approximately 2,000 people in the district.

The second-smallest (next to Salmon Point) of the ES districts, the school population in Oyster Cover was 286 students, served by a staff of 27 in two schools. At the time of the ES competition the annual school budget was $321,000 and the per-pupil expenditure was $1,016.

In Oyster Cover the schools are supported by a combination of property-tax revenue and annual special levies. These levies cause periodic scrutiny of the schools and some criticism of their expenditures and quality of education. In 1970–1971 a concerned-citizens group prevailed upon the school board to request an external evaluation of the system by the state. The letter of interest prepared by the superintendent was based on the recommendations made in that evaluation.

As in Liberty Notch, the requirements of the ES program for a K–12 population to be involved in any local project caused a partnership with the neighboring community, which sends its eighth-grade graduates to a high school in Oyster Cove but is otherwise autonomous. This temporary partnership was a source of some problems in planning and implementing the project. In 1974, the neighboring community dropped out of the program and Oyster Cove continued alone.

Teachers in Oyster Cove are somewhat younger than in the other district, and the district experiences a rather high staff-turnover rate.

TIMBER RIVER

Timber River school district is located 250 miles south of Portland, Oregon. Its population of approximately 10,000 is predominantly white, although there are some American Indians. Employment is diverse and includes lumbering, mining, farming, tourism, and business.

The school system at the time of entry into ES contained six schools with a total enrollment of approximately 2,500 students served by a staff of 118. The annual school budget was $2,223,848 with a per-pupil expenditure of $884. In terms of assessed valuation per pupil, Timber River was the poorest district in its state.

In its letter of interest, Timber River focused on eight top-level administrators as the district's major strength. Although this was a rather unusual statement (other districts tended to include teachers and community residents as a strength), Timber River did have an unusually sophisticated and experienced administrative staff; for instance, the superintendent and associate superintendent had both formerly been federal program officers in Washington, D.C.

The letter portrayed Timber River as a district which had already taken several steps prerequisite to educational change, including needs assessment, staff training, and some curriculum-development efforts, and proposed the creation of a "new learning climate" throughout the school system.

DESERT BASIN

Desert Basin is situated in the southeastern corner of Arizona, 85 miles from Tucson. In 1970, the school-district population was approximately 4,500, with a sizable Mexican-American population. Forty-eight percent of the county population is involved in white-collar occupations of various types.

The school system had the longest history of being consolidated of any of the ES districts (32 years), and served 1,433 students at the time of the district's entry into the ES program, of whom 26% were Mexican-American. There were three schools located on a single campus—an elementary school, middle school, and high school with a total staff of 78. The 1971–1972 budget was $1,233,120, with an average per-pupil expenditure of $886.

The letter of interest stated that the bicultural heritage of the residents was one of the strengths of Desert Basin, and that elements of cultural diversity should be built into the curriculum of the schools. The major problems of the school system, according to the letter, were almost all in the area of curriculum, which was said to lack both scope and sequence. The secondary focus of the letter was on the community: finding ways to draw in community members as a cultural resource (particularly the Mexican-Americans) and also finding ways to make use of the school facilities as a community resource.

The Assessment of Readiness

These 10 districts were the "winners" of the ES program's selection process. However, our initial assessment of the conditions in each district suggested that the districts differed notably in their readiness to embark on a complex process of planned change.[7] These differences, when combined with insights gleaned from the literature on organizational change, provide important clues that will be helpful in understanding why the 10 districts varied in the effectiveness which they demonstrated in the subsequent stages in the change process (see Chapter 9).

We have identified four factors which seem to have played a major role in the apparent readiness of the districts to take advantage of the opportunity presented to them in the ES program.

1. *The degree to which social forces external to the school system were*

[7]The data sources of this assessment include the letters of interest, historical accounts of each district written by trained social scientists who were assigned by Abt Associates to live full-time in each school district for at least three years (Fitzsimmons *et al.*, 1975), periodic interviews with each of these "on-site researchers," and questionnaires completed periodically by professional personnel in each district.

impinging on it to change. Quite frequently, social and cultural changes occurring in the external environment of the organization create pressures for change (Baldridge & Deal, 1975; Iannaccone & Lutz, 1970). During the period prior to March 1972, many of these small, rural school districts were experiencing such pressures from their environments. In Clayville, for example, expansion in the local economy had recently occurred; in Liberty Notch it was a case of gradual economic decline. In Big Sky, recent changes in state law had necessitated school-district consolidation. In Magnolia, racial desegregation had recently been achieved. In Oyster Cove, several bond issues had recently been rejected by the citizens.

In assessing the readiness of each school district, we assumed that those such as Clayville whose sociocultural environments had undergone recent or dramatic change were more ready to embark upon a program of comprehensive change of the type envisioned by the ES program than those such as Liberty Notch which had experienced more distant or less dramatic environment change. This assumption is based on the capacity of sudden or dramatic change to "unfreeze" the system and thus make it more receptive (or more vulnerable) to further change.

2. *The degree to which there was a formal recognition within the school system of unmet educational needs.* Just as these 10 school districts varied in the degree to which they were experiencing pressures for change from their environments, they also differed in their recognition of unmet educational needs. Such a recognition can come in a variety of forms: a pronouncement from the superintendent, the demands of a teacher organization during a process of collective bargaining, editorials in the student newspaper. However, those statements which are based upon a formal process of needs assessment, conducted by experts and presented in writing, seem to acquire the greatest legitimacy, both within the school system and within its resource environment.

There are many motivations for the preparation of such statements. All regional accrediting associations require the periodic preparation of "self-studies" as a part of the process of accrediting high schools. Many states require such studies of the entire school system. Occasionally, dissident parent or teacher groups will conduct such studies, which, although initiated outside the formal boundaries of the school system, are often subsequently accepted by it as legitimate statements of need.

In assessing readiness of each school district, we assumed that those such as Clayville which had engaged in a formal needs-assessment process during the five years immediately prior to 1972 were more prepared for a program of comprehensive change of the type envisioned by the ES program than those such as Big Sky which had undergone a less formal or less recent assessment.

3. *The degree to which there was broad-based experience within the school system with competitive federal programs of educational improvement.* During the 1960s, a series of federal programs was inaugurated which made federal funds available to public school districts for the improvement of their educational programs. Some of these funds (e.g., ESEA Title I) were given as general grants to any district which could qualify on the basis of the characteristics of its student population. Other funds (e.g., ESEA Title III) were more categorical and available only to support demonstration innovative projects or innovative programs in competitively selected districts. Although all experiences with federal funding may increase readiness, those involving competitive selection for innovative projects seem especially salient.

In assessing the readiness of each school district for externally induced change, we assumed that those such as Salmon Point which had the greatest prior experiences with competitive federal funding were more ready to embark upon a program of comprehensive change of the type envisioned by the ES program than those such as Butte–Angels Camp which had a lesser degree of such experience.

4. *The degree to which the administrative leadership of the school system was committed to the type of change envisioned by the ES program.* Although external social forces, internal recognition of unmet needs, and past experiences with competitive federal funding all seem to be important theoretically in establishing the readiness of a school district to begin a process of externally induced planned change, there is at least one additional factor which seems especially important: the commitment of the administrative leadership of the district to the innovative endeavor.[8] Although the responsibility for implementing planned changes eventually necessitates wider involvement, at the readiness

[8]It must be noted that commitment to the innovate endeavor may be based primarily on "opportunistic" motives or "problem-solving" motives (Greenwood *et al.*, 1975). However, it is beyond contention that in chronically underfunded rural school districts the two motives interact, and that the notion of administrative commitment is more important than the motive for that commitment.

stage administrative leaders are likely to be of prime importance. If, as in Timber River, more than one of the key administrators were actively involved in the process of assessing internal needs and in matching such needs with external realities, the school system would seem to be more likely to move ahead. If, as in Liberty Notch, the initiative for seeking external resources had come from lower levels of the system (e.g., a building principal or a teacher representing a particular subject-matter area), and had not been rationalized on a system-wide basis by the direct involvement of the administrative leadership of the entire system, the change effort would be more easily undermined at subsequent stages.

In assessing the readiness of each school district we assumed that those such as Prairie Mills in which the superintendent of schools led a system-wide team of other district-level administrators and building principals (when such positions existed) in the preparation of the letter of interest were more ready to embark upon a program of planned change of the type envisioned by the ES program than those in which this was not the case.

Table 4. Readiness Profiles for the 10 School Districts

	Readiness factor			
School district[a]	External social forces	Internal recognition of unmet educational needs	Previous experience with competitive federal funding	Commitment of administrative leadership
Prairie Mills	M	H	H	H
Magnolia	M	H	H	L
Salmon Point	L	M	H	H
Timber River	M	M	M	M
Clayville	H	H	L	M
Oyster Cove	M	M	M	M
Butte–Angels Camp	M	H	L	M
Desert Basin	M	M	M	M
Big Sky	H	L	M	L
Liberty Notch	L	L	L	L

[a]The order of school districts corresponds to their overall readiness score. *Key:* H = high, M = moderate, L = low.

A Summary of Readiness

The descriptions of the 10 selected school districts presented earlier portray their diversity on many factors. This diversity becomes even more apparent when we examine each district's distinct *profile of readiness* based on the four readiness factors (see Table 4). As described in Chapter 2, the rational approach to the study of change posits that "success" at earlier stages affects success at subsequent stages in the change process. In order to test this notion we computed a single "best esimate" of the readiness of each of the school districts, based upon some simplifying assumptions about the degree of readiness represented. We assigned the numerical scores of 3, 2, and 1 to the categories of "high," "moderate," and "low," respectively, for each factor of readiness.[9] We then computed an overall "readiness" score for each district by adding together its scores on each of the factors.

Thus, Prairie Mills is estimated to have had the highest degree of readiness (11 units). It obtained this score by receiving 2 units for being moderate in external social forces and 3 units each for high commitments of administative leadership, internal recognition of unmet educational needs, and previous experience with competitive federal funding. Liberty Notch is estimated to have had the lowest degree of readiness (4 units), obtained by being low on all four factors. The

[9]The assignment of districts to these categories was based on a set of criteria appropriate to each factor. Using as an example "previous experience with competitive federal funding," the criteria were:

High: Included in this group are districts with a history of seeking and acquiring competitive funds and a favorable experience with that process.

Moderate: Included in this category are districts with a more limited history and those in which the superintendent may have had such experience in a previous location even though the district itself has had none.

Low: Included in this category are districts which had had no previous experience with competitive federal funding or in which such experiences had been unfavorable.

Data for assignment of districts to categories were derived from historical accounts of each school district and from interviews with the OSRs. Each district was ranked independently by three members of the project research staff. Consensus was then reached by rating the districts in three categories—high, medium, and low—for each factor. It is important to note that our approach makes a series of "simplifying assumptions." We recognize that other plausible assumptions could be made which might lead to a different classification of the 10 school districts in terms of their overall readiness.

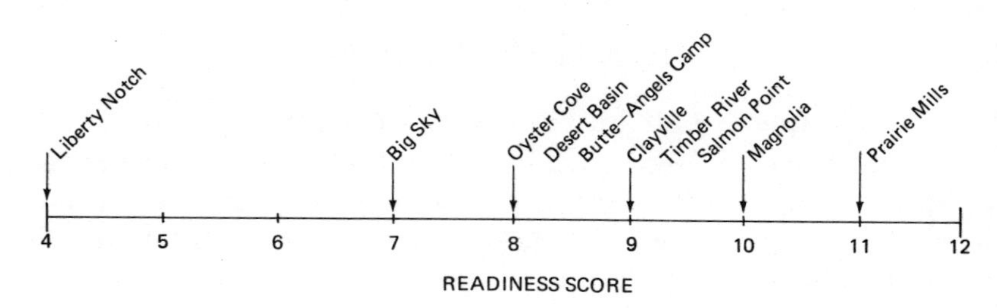

Figure 2. Readiness scores for the 10 school districts.

results of this scoring procedure are presented in Figure 2, and its predictive utility is considered in Chapter 8.

This chapter has introduced the partners in a large-scale effort at planned change, first by presenting the assumptions of the federal planners and then by describing the 10 participating districts and their readiness for change. The federal-local relationship developed and deepened during the year of planning that followed the districts' entry into the program. Chapter 5 traces the course of the relationship by giving an account of the expectations, activities, and outcomes of that year.

5

The Beginning of Change

The Planning Year

The kind of partnership envisioned by the ES program designers was described by one subsequent observer as requiring "the creation of a trust and confidence . . . built on perceptions of mutuality, equal risk for equal gain, respect and caring" (Gideonse, 1979, p. 438). In some respects, the federal-local relationship reached its highest point of satisfaction for both partners in June 1972, when planning grants were awarded to the "winning" rural school districts, and the planning year, the initiation stage, was about to begin.

When two administrators and one school-board member from Big Sky returned home from their first meeting with ES/Washington (immediately following the award of the planning grant), one administrator wrote the following comment to the board:

> We were pleasantly surprised to find that there are no guidelines and that in this project we will be able to do what we want to do, not what someone else wants us to do. (quoted in Messerschmidt, 1979, p. 55)

The superintendent and one principal from Desert Basin returned home from that meeting in a similar state of elation, convinced that ES/Washington was as interested in and enthusiastic about the local project as they were (A. Burns, 1979, p. 97).

However, as many events of the planning year illustrate, the partnership came to be frequently characterized by misunderstandings, misperceptions, and distrust. And at times, ES/Washington officials and local educators appeared to be adversaries, rather than collaborators, in an important educational enterprise (Gross & Herriott, 1979).

The inability to maintain the generally pleasant initial relationship can be attributed in part to the loose linkages that existed between the federal agency and the local districts. For example, the stated goals of ES/Washington were generally not the same as those of the local districts. Nor were the goals of all the districts or their motives for participating in the program similar. Federal program planners apparently designed the program to attract school districts with strong problem-solving motives. They built into the program the opportunity for local problem identification and solution. However, they also built into the program elements which held strong attraction for "opportunistic" participation (Greewood *et al.*, 1975). These included the very large amount of money to be awarded, long-term guaranteed funding (apparently without the need to reapply), and the impression conveyed by the Announcement of "no strings attached," that is, little accountability in use of funds.

The impacts of culture barriers between the urban, bureaucratic Washington agency and the small rural districts with which they were dealing emerge again and again in case studies (see, e.g., Herriott & Gross, 1979, Chapters 3–8) and in Corwin's (1977) study of the ES/Washington program officers. This problem took many shapes and was more acute in the more rural and less experienced districts. As simple a matter as the name of the program—Experimental Schools—although important to ES/Washington in selling the program within the federal bureaucracy, was a source of confusion and discomfort to the rural districts. There was a perception at the community level that it implied experimenting with their children; this evoked fear and hostility. Some districts (Desert Basin, for example) did not refer to the program locally as "Experimental Schools," but as the "Small Schools" and "Rural Schools" project.

Loose linkages between the two levels were loosened still further as the years passed, in part because of the state of affairs at the federal level. Frequent turnover in program officers whose role it was to monitor and interact with the local school-district personnel, as well as erratic and irregular communication between the program officers and the local districts, were especially problematic. In many instances, the planning year was characterized by excessive interaction and control by the federal program office, whereas in later years the lack of communication and guidance was seen by some local personnel as a sign of disinterest in their problems. What emerged was an uncomfortable partership between the federal and local levels, one in which it was never clear which "partner" was to control the relationship (Corwin, 1977).

FEDERAL EXPECTATIONS FOR THE PLANNING PERIOD

The federal assumptions behind the ES program (described in Chapter 4) aid in understanding the basis of the federal-local relationship. Similarly, ES/Washington's expectations for the local districts during the planning period can be examined as the framework for the federal-local relationship and its ensuring outcomes as the projects were negotiated during the planning year.

Although it was generally not clear to the administrative leaders of the 10 schools districts at the time the planning grants were awarded, the staff of the ES program expected to play a very active role during the initation stage of these projects. There seemed to be a sense within ES/Washington that the anticipated budget for the undertaking (over $8 million), and the fact that the districts would subsequently be awarded four-year contracts, made it essential for the federal government to ensure successful launching of the projects.

The expectations which ES/Washington had for these 10 small rural school districts during the planning year were communicated in several forms. A formal grant document was prepared by the ES program in June 1972 and subsequently accepted by the board of education within each district. The document made explicit the number of dollars available to the district and discussed more generally the purposes for which those dollars could be spent. A set of guidelines for the preparation of a formal project plan was also prepared by the ES program. Finally, a federal monitor (referred to as the ES/Washington Program Officer) was assigned to oversee the grant relationship between the federal government and each school district. Such monitoring took a variety of forms and occurred through visits by the staff members and consultants of the ES program to the school districts, visits of school-district staff members to Washington, D.C., periodic telephone calls, letters, and memorandums.

Although ES/Washington staff members frequently said different things to their counterparts in different school districts, eight general expectations applied to all the districts.[1] These were

1. Production of an acceptable formal project plan
2. Critical reexamination of assumptions in letters of interest
3. Broad participation by affected groups
4. Prior approval by ES/Washington of budget and staff
5. An interactive process of plan review and revision
6. A commitment to comprehensive change
7. Formative evaluation of project progress
8. Contracting for plan implementation

The descriptions which follow point out what some of these expectations implied for particular school districts.[2]

1. Production of an acceptable formal project plan. In order to be eligible for long-term funding, each school district was expected to produce a formal project plan. The plan was to be no more than 100 pages long. It was required to be acceptable to the federal government as a local statement of comprehensive goals for the school district, and of the procedures and resources necessary for the accomplishment of those goals. It was the intention of the ES program to contract formally with each school district for the implementation of the changes spelled out in that plan. Achievement of this goal, however, implied the district's prior accomplishment of ES/Washington's other expectations.

2. Critical reexamination of the letters of interest. There appears to have been an attempt on the part of ES/Washington to counteract the "grantsmanship" aspect of the letters of interest and to encourage the districts to look at their needs realistically as the foundation for putting together a useful and implementable project plan. A careful needs assessment on the part of the local district was required—a needs assessment which could not have been easily conducted before the

[1]This compilation of federal expectations for the planning year was first presented in Herriott and Rosenblum (1976) and further expanded in Herriott (1979).

[2]These descriptions draw heavily from the site case studies prepared by the OSRs and from Corwin (1977), and are referenced where appropriate within the text.

letter of interest was written because of the short time (four weeks) allotted. How did the districts seek to fulfill this expectation? In general, the opportunity to back away from the ideas generated for the letter of interest was not taken by the districts; instead, they tried to operationalize them.

In Clayville, the superindentent had responded to the ES opportunity eagerly, seeing it as a means to upgrade the district and to enhance his own career. Much of Clayville's letter of interest had been based on a systematic needs assessment recently carried out in the district to see if schools were a problem and if they were using all taxing authority. The univeristy-based consultants hired by the superintendent to write the letter of interest focused on the concept of diagnostic instruction. This concept was embraced by the superintendent as a means of "winning" the competition; as remarked, "It's the in phase." There is no indication that the superintendent ever reexamined the idea of diagnostic instruction, the basis of the formal project plan, in view of the actual needs and priorities of Clayville (Clinton, 1979a, p. 222).

Magnolia's letter of interest was written by one person, a newcomer to the district who was serving as curriculum director and was familiar with the federal proposal process. Her letter called for a "continuous progress, individualized non-graded program" in reading. Although she had not had time to solicit ideas for inclusion in the letter from either school personnel or community citizens, and although she herself was very soon to leave the district, there is no indication that any reexamination of the direction taken by the letter of interest took place (Wacaster, 1979).

The needs-assessment process was approached in several different ways by the districts. Perhaps the most common was to hold meetings with teachers, citizens, and community groups in order to develop an informed assessment of problems and needs. These meetings showed a tendency to be unfocused and to end up as "gripe sessions." Some districts, however, attempted more formal needs assessments, but with poor results. For example, ES/Washington suggested that Big Sky obtain help from a nearby university in conducting a formal needs assessment. At first, the superintendent was enthusiastic, but when he and his assistant met with university personnel they found out that it would take an entire year to conduct the kind of study the consultants wanted to do. The local administrators decided against using "outsiders" in the project (Messerschmidt, 1979).

Liberty Notch, a loose confederation of three small, autonomous districts that were required to join together for the ES project, also found the experience of using "outsiders" to be less than fruitful. Uneasy in the relationship and unused to planning together, they turned to an urban-based consulting firm which offered to carry out a needs assessment for the district using a questionnaire tailored and personalized to the district. When the questionnaire was shown to the local committee, they judged that it was not personalized to the district and that changes should be made in it. However, the consultant replied that the questionnaire could not be altered. The climate became hostile and the data was consequently never used (Stannard, 1979).

3. *Broad participation of affected group.* This expectation required expanding the planning group beyond the few who had participated in preparing the letter of interest to include principals, teachers, and community members. The requirement for input from all these groups in shaping the goals and objectives of the local projects became a source of conflict within the district and between the districts and ES/Washington. Nevertheless, all the districts attempted to fulfill this expectation in one way or another.

The superintendent in Timber River began the school year (1972–1973) by forming teacher committees to write "proposals" for inclusion in the plan. An ES coordinating council was set up, which included prominent residents, teachers, and three high-school students. Hundreds of meetings were held to consider the proposals submitted by faculty, who were given release time for writing. Working under tight deadlines, the administration then prepared a rough draft of the plan to submit to Washington. Apparently to demonstrate their professionalism, the administrators chose to phrase the document in highly technical terms; the result was unacceptable to the local staff. When the teachers saw the document, their reaction was that it did not represent their concerns and that their efforts had been merely a "smoke screen." The net effect of Timber River's broad participation appears to have been negative—many staff members felt disenfranchised (Hennigh, 1978).

Magnolia took quite a different approach to achieve participation. The superintendent began by delegating to principals the responsibility of forming school-based planning committess, which typically included the principal, members of the school board, a district board of education member, teachers, and parents. ES/Washington was dissatisfied with so fragmented an approach and suggested that a *district-wide* coordinating committee be formed to include representation from all these school-based committtees. The superintendent formed such a committee and informed ES/Washington that it would incorporate the school-based plans into a district-wide plan. (However, several principals were upset to hear that there would not be individual school projects.) A consultant retained by the superintendent wrote the plan, in which he developed a suggested theme and listed 38 goals. The plan was mailed to Washington exactly as prepared by the consultant, but with the individual school plans appended.

ES/Washington responded that this submission was unacceptable and lacked comprehensiveness. Despite additional consultant help, Magnolia's second attempt proved no more acceptable and ES/Washington decided that Magnolia was not able to produce a plan on its own. On her next visit, the program officer brought with her an education consultant from Florida who she suggested could help with revising the plan. The final plan was prepared by the consultant with information funneled to him by the principals (Wacaster, 1979).

4. *Prior approval by ES/Washington of budget and staff.* Approval was required by ES/Washington for the appointment of staff in major roles in the local projects and for anticipated expenditures for equipment in any amount over $200. ES/Washington considered this to be a reasonable expectation for a federally funded project. However, to many of the districts, this was a new and frequently unwelcome process, and cast doubt on their "no strings attached" perception of the program.

Once again, Magnolia experienced the kind of authority ES/Washington could exert in project staffing. Magnolia's project director was originally chosen by the superintendent primarily because he was a long-time resident, teacher, and administrator in the system, and because he could "get along" with everybody. ES/Washington, however, took a dim view of the appointment and eventually demanded that he be removed because he did not appear competent to direct and implement the project (Wacaster, 1979).

If ES/Washington could remove staff, they could also install them. For example, Butte–Angels Camp presented a plan to ES/Washington in which their specifications for the required evaluation plans were not clear. Since the district had no staff with evaluation expertise, ES/

Washington insisted that a consultant designated by ES/Washington be used in that capacity (Firestone, 1979).

Nor did the disbursement process always run smoothly during the planning year. Because of the transfer of the program from OE to NIE and the resultant bureaucratic snags at the federal level, and because of the turnover in program officers assigned to monitor some of the districts, there were long delays in payment and approval from ES/Washington. Since all the districts were in straitened financial circumstances, they did not have the slack in resources to tolerate these delays which, consequently, caused frustration and hostility. Not only were there snarls in specific budget items but in overall payments of ES funds to the districts as well. Since the districts were geneally unfamiliar with the bureaucratic procedures at the federal level, they had no basis for understanding what often appeared to be capricious behavior by ES/Washington. For instance, the superintendent in Desert Basin reacted angrily when ES/Washington failed to send reimbursement to the district for the previous quarter:

> they do things like this all the time at their own sweet time. They're up there in Washington, all on vacation. So because they go on vacation, our people go without being paid. They think "the hell with little districts." (quoted in A. Burns, 1979, p. 205)

5. *Interactive process of plan review and revision.* There seemed to be an implicit assumption on the part of the staff of the ES program that small rural school districts would have difficulty in producing an acceptable plan without periodic review of written drafts by the ES staff and their subsequent revision by the planners. The guidelines called for the submission of a preliminary draft of the formal project plan by December 15, 1972. For many of the 10 school districts this was merely the first step in what subsequently proved to be a long process of review and revision, continuing in some instance well into the implementation year.

The approaches taken at the district level to preparing drafts varied greatly, as did the kind and amount of ES/Washington involvement in the preparation. Sooner or later, however, each district came into conflict with ES/Washington over some element of its formal project plan.

Prairie Mills, for instance, prepared its first draft with the help of consultants and submitted it to ES/Washington in December 1972. In January, ES/Washington replied that the plan was overly ambitious and would require substantial reorganization. "Overly ambitious" conveyed to some Prairie Mills personnel an undertone of prejudice against the capabilities of rural school systems, and this was resented. Local project personnel were also disappointed that no directions were given as to how to revise the plan. Written in uncertainty, the second draft was sent off, but it too was termed "unacceptable" by ES/Washington. Prairie Mills was informed that a "complete rewrite would be necessary" (Donnelly, 1979b, p. 198).

The supervisor and the ES project coordinator in Prairie Mills became angrily convinced that local initiative was no longer a Washington priority, that by adjusting the plan to meet ES/Washington expectations they would make it less acceptable to local residents, and that they had neither the time nor the money to devote to the task. ES/Washington became aware of these local feelings and in April arranged for a consultant to assist in completing the plan. His first advice to the project leaders was to concentrate on the five components of the plan which were planned most thoroughly and delay implementation of eight others. The project leaders replied that they were fully capable of activating all aspects of their plan—all they needed was help in preparing an acceptable final draft of the plan. Finally convinced that they were right, the consultant told ES/Washington that the Prairie Mills plan should be accepted. Prairie Mills might be viewed as a district which took a stand for local autonomy with some success, for during the first three years of implementation, ES/Washington presented a relatively low profile to the district (Donnelly, 1979).

A very different story is that of Desert Basin. The same small core group which had written the letter of interest from Desert Basin continued to be the major actors during the planning year. This group remained in close contact with ES/Washington. The relationship between the local and the Washington ES personnel was one of "partnership" and included weekly telephone calls, correspondence, and site visits. When the final plan writing began in January, an outline format was sent by ES/Washington which was closely followed in the writing. Only minor revisions were required. However, by the time the plan was accepted, all the core group members had left the district. Many residents, teachers, and administrators felt that this group had "owned" the ES project and should have taken it with them. One resident remarked, "We got rid of the last superintendent, and when we were rid of him, we thought we were rid of the Rural Schools thing too" (A. Burns, 1979, p. 115).

Clayville exemplifies still another kind of planning process. Its superintendent had hired consultants to write the letter of interest. He attempted to use much the same style in preparing the formal project plan, only to encounter a negative response. ES/Washington asked, "When and how does the public get to *influence* the status of the project?" and went on to criticize the overuse of consultant help (Clinton, 1979b, p. 224). The superintendent, who believed he must comply with the wishes of ES/Washington, was forced to send his consultants back to the university and embark on a new draft without the expertise he depended on. This draft, which took a great toll on his staff and was not completed until the early morning of the day it was due, was also received negatively. Although Clayville was eventually able to produce a plan acceptable to ES/Washington, the superintendent and the two ES coordinators believed that "impossible demands" had been put on them by ES/Washington (Clinton, 1979b).

The euphoria with which the districts received notification of their planning grants was based on their belief that these funds would allow them to plan their own projects locally and independently of federal control. The review and revision process was therefore particularly disillusioning.

6. Commitment to comprehensive change. This expectation was perhaps the most difficult aspect of ES project planning for all the rural districts and the one which was the direct cause of many of the revisions of the project plans referred to above. No written definition expanding on the concept of comprehensiveness that was articulated in the Announcement (i.e., the facets of change) was made available to the districts, and very little guidance (in the eyes of the local personnel) was offered from ES/Washington.

Some of the ES program officers working directly with the sites were also confused about comprehensiveness. One stated, "I was responsible for going over their drafts and criticizing them from some basis which wasn't really specified . . . something to do with 'feasibility' and whatever 'comprehensiveness' was" (Corwin, 1977, p. 17).

Each Rural ES district has a "comprehensiveness story" in the annals of its planning year. Big Sky's illustrates the direct impact of this expectation on the content of the local project. Big Sky was primarily interested in ES participation in order to help unify its recently consolidated district. The planning of its project was to be an important aspect of bringing the district together—at the administrative level at least. It is for that reason that the first central theme arrived at by Big Sky planning personnel was a "statement of coordination." Such a theme was not acceptable to the federal program officer, who had suggested one which emphasized individualized instruction. She called Big Sky's approach a "statement of a method of coordination," not a theme, and felt that its provisions for comprehensive change were unclear (Messerschmidt, 1978, p. 123). The planning committee (under pressure) agreed to a compromise theme, "the development of a community involved education program that emphasized a future-oriented learning environment for each student and his individual needs" (Messerschmidt, 1979a, p. 104).

On March 30, 1973, the Big Sky plan containing this theme was sent off to Washington, and not until final budget negotiations, which began in May, did the planning committee find out that ES/Washington wanted extensive revisions to it. A memo received in early July contained the scope of work for revisions, including a theme devised by ES/Washington for Big Sky's ES project. It stated:

> The Project in [Big Sky] will include the entire school district in a systematic process of change destined to result in the delivery of educational services which will prepare students for productive work and enjoyable leisure in accord with their individual needs and interests. All aspects of the educational system will be coordinated in such a way as to support the integration of themes and skills of career education and cultural education with ongoing curricular efforts. Community-participation will be encouraged throughout all phases of the educational process. (quoted in Messerschmidt, 1972a, p. 121)

This unexpected setback and other controversies within the district and between the district and ES/Washington led to discussion in both Big Sky and ES/Washington of canceling the project. The resolution of the situation was the eventual appointment of a new program officer and a quiet disregard of the explicit theme(s) of the project.

7. *Formative evaluation of project progress.* This expectation was written into the Announcement and required as part of an acceptable project plan. The intention on the part of ES/Washington was that districts develop a capability for judging systematically the success/failure of each local project "in terms of its own purpose and goals" (The Announcement, p. 4). Such "formative evaluation" was a difficult process for most of the districts to grasp and very few had staff

with expertise in this area. Indeed, this particular expectation was to cause problems for the districts not only in the planning year but throughout implementation as well.

In its attempt to complete an acceptable project plan, Big Sky contracted with an expert to conceptualize and write a section on evaluation; he did this quite independent of local officials. Despite the written evaluation design, implementation lagged as Big Sky personnel still thought in terms of traditional evaluation, that is, informal, subjective, and comfortable. A change from this traditional method could be accomplished only by pressure from outside the system. Implementation began only when, along with the insistence of the ES/Washington program officer, a new state law was added which demanded systematic evaluation by all Local Education Agencies (LEAs) within the state (Messerschmidt, 1979).

Even where attempts were made to implement evaluation as planned, things did not always proceed smoothly. Oyster Cove had set up a faculty project-evaluation team at the suggestion of their federal program officer. The faculty members felt unable to do the tasks and asked the program officer for help. In his desire not to influence the "local autonomy" of the project, his answers seemed to be less than helpful. The committee chairman felt "he knew what to do," but wouldn't tell them. Instead of answers the committee chairman was given a book on statistics; she later commented that she now knew how to calculate a standard deviation but did not know what it was or how it applied to the evaluation process (Colfer & Colfer, 1977, p. 194).

8. Contracting for plan implementation. A formal project plan was to be the outcome of the planning year; the plan would then serve as the basis of the contract of implementation to be signed by each district and by NIE. A preliminary draft was to be submitted to ES/Washington by December 15, 1972, and a final product by April 15, 1973. For the districts which had not yet been given moral commitments of continued funding, this remained a period of uncertainty.

Two major points underlie the districts' misunderstanding of this important expectation and the problems which resulted. First, the districts did not all understand the ES/Washington concept of contract negotiations. Second, they did not understand what function ES/Washington intended the formal project plans to serve. Two districts

(Butte–Angels Camp and Big Sky) found contract negotiations particulary grueling.

ES project personnel in Butte–Angels Camp had experienced a difficult planning year, but by the time they reached the point of negotiating with ES/Washington they believed the worst was over. On May 31, 1973, the superintendent, a counselor, an elementary teacher, and a school-board member arrived at ES/Washington's home office to "negotiate" the final contract. District personnel expected a routine meeting to clarify details and fix the final dollar amount. Instead, they underwent a harrowing experience. At one point during the first day, the ES program director's abrasive questioning of the school-board member about technical jargon buried in the district's plan reduced her to tears. The superintendent and counselor sat up most of that night trying to revise the plan. They were left with a fear of ES/Washington that lasted until a new ES/Washington program director was named two years later (Firestone, 1979).

Big Sky also misunderstood the contract negotiation process. When the superintendent, the temporary planning director, and the incoming assistant superintendent came to negotiate in Washington, they believed they would negotiate a contract for $596,000 with NIE for the full *five-year* period. Instead, they were offered a *three-year* contract at $361,351 (approximately 60% of the total amount), and were told that the two final years of funding would depend on renegotiations in 1976.

> What was apparently standard procedure in Washington—negotiating a contract in 2 parts, 60% funding for the first 3 years and the remaining 40% later—was interpreted by Big Sky personnel was an unwarranted "cut" of 40% overall. (Messerschmidt, 1979b, pp. 92–93)

They believed the meetings had been controlled by NIE and that they had little voice in the decisions; they returned to Big Sky suspicious and cynical.

The belief in ES/Washington as to the purpose of formal project plans was that they would serve as the blueprint for implementation. When NIE negotiated a contract with each district, they saw themselves contracting for "delivery" of the locally planned project components. The understanding on the part of ES/Washington of the federal role in the contract was expressed by a program officer: "Contracts spell out pretty closely what these people are going to do by such and such a time, and if they're not delivering, then you can step in" (Corwin, 1977, p. 49).

This understanding of the basis of a contracted relationship was somewhat unilateral. It was the perception of many local project planners that these plans were a stepping stone into the project and that they could be altered, components of them discounted or new ones instituted at local discretion. The change from one-year planning grants to three-year implementation contracts proved a generally confusing one.

THE PRODUCTS OF THE PLANNING PROCESS: AN OVERVIEW OF THE ES PROGRAM

At the end of the planning year, the 10 school districts signed contracts with the ES program in Washington for the implementation of their projects as described in their formal project plans. Each project was subdivided into a set of operational components which had been grouped together around a unified theme to affect all facets of educational change in a mutually supportive way. Each district identified project components in its own way, some at a less differentiated level (e.g., "basic skills"), some at a more differentiated level (e.g., "reading"). Although any of these components, taken alone, might be considered a piecemeal endeavor, all of them together would, it was hoped, affect all facets of the organization's functioning.

Overall, the 10 school districts committed themselves to implementing a total of 146 components (only four of the 146 planned project components were never implemented in any form, though others were discontinued during the course of the projects.) Required by ES/Washington to make changes in all five facets of schooling in the aggregate, they had planned 73 components to change curriculum and 46 components to change and improve their instructional staffs. Altered use of time, space, and facilities was involved in 57 of the components. Forty-six components were designed to involve the community in various degrees of active participation, and 34 changes were proposed which touched upon the administration and organization of the school districts.[3]

[3]The components included under each facet add up to more than 146 because 80% of the 146 project components affected more than one facet.

Typical Project Components

Curriculum

Central to many of the projects, and probably of paramount interest to many of the local planners, were changes in curriculum. Four major areas of curriculum were addressed by components in all 10 projects: career education, outdoor education, cultural enrichment, and basic skills. Individualized-instruction technology, permitting each child to gain skills in a subject at his or her own pace rather than as part of a group or a class, was introduced in at least six of the projects. Career education often brought local businessmen into the classroom to lecture, or made it possible for students to receive on-the-job training. Some districts used their rural environments to provide instruction in ecology, biology, and outdoor-survival skills. Cultural enrichment often involved expanding already existing arts-and-crafts and music curricula.

Staffing

One of the stipulations of ES/Washington was that support structures be provided for teachers and staff undertaking new instructional methods and new curriculum options. All 10 projects, therefore, held in-service workshops to train teachers in using new curricula. Within this facet fell summer training sessions; during the year, release time allowed teachers to visit other classrooms, other schools, and even other districts. Provisions for teachers' aides, clerical help, and media resource centers were included in six of the 10 projects.

Time, Space, and Facilities

In all the projects, alteration of the use of time, space, and school facilities was an adjunct to changes in curricula. For example, school activities such as field trips were scheduled during weekends and vacation periods to increase the amount of instructional time; on-the-job training and environmental education also expanded the use of space by taking education outside traditional school boundaries. Media and resource centers permitted a more flexible arrangement of space than the traditional self-contained classroom. School facilities were made available to the community for cultural events.

Community Participation

To a greater or lesser degree, the 10 school districts all planned and implemented citizen advisory committees. The project also involved local residents in the schools as volunteers in four ways: as aides in classrooms, as supervisors in the new vocational and environmental programs, as chaperones for field trips and camping programs, and as career counselors in proposed vocational guidance programs. Adult education was another way in which community participation was increased.

Administration and Organization

Components affecting this facet concerned either advisory or decision-making functions. Advisory bodies were set up in all the districts. Changes in decision making were not as prevalent, although two districts set up citizen bodies with some power over decision making. In all the districts, new roles were established for the support and management of the ES projects.

In contrast to these popular approaches, the components planned and implemented by some districts represent unique responses to common problems. For instance, all 10 districts were concerned with career and vocational education, but Prairie Mills provided a practical auto-mechanics course in which students used machine parts and car engines donated by local merchants. In Liberty Notch, students actually constructed a house. (This project's intern program also provided volunteers to a local hospital.)

Among the many innovations stressed in ES projects, very few involved change in the traditional role of "students as learners," even when individualized instruction and flexible schedules were implemented. A component in Prairie Mills, however, attempted some change in that area by giving older students the opportunity to tutor younger ones. This district also used students as teachers' aides in the classrooms.

The problem of cultural isolation was dealt with in components which brought the outside world to the students and sent the students to the outside world. Big Sky, for instance, arranged for visits by poets, artists, and musicians for its students and citizens, while Salmon Point gave its students opportunities for travel experience beyond the borders of the continent and even to China and the Soviet Union.

Another problem which plagues most rural school districts is that students must travel long distances to school. The time rural students spend on the school bus and the nature of the busing schedules has traditionally made it difficult to use before- or after-school hours for extracurricular activities. The ES project in Prairie Mills attempted to solve the problem of extensive travel time in several ways. In three school buses audiovisual equipment was installed to make use of travel time for foreign-language practice. In addition, shuttle runs supplemented the regular bus schedules to allow flexibility in planning extracurricular activities after school.

Content of the Ten ES Projects

Although they were intended to address local needs and problems, there is a remarkable similarity in plans for the 10 ES districts, reflecting the needs and problems common to all 10 school districts, their awareness of educational trends being advocated in the popular educational literature at that time, and the process of review and revision imposed by the staff of ES/Washington. However, each project was also unique in its content, in the way in which the project was organized, in the number of components planned, and in the strategies chosen for implementation. We present very brief descriptions of each of the ES projects, highlighting some of their components. For each district, we give the number of components in operation at the time of our analysis of implementation (the focus of Chapters 6 and 7). In addition to the activities mentioned in the capsule descriptions, all projects planned and implemented staff training activities and established community advisory committees. (More detailed descriptions of all 10 projects appear in Appendix A.)

SALMON POINT (NINE COMPONENTS)

A district chronically afflicted with high seasonal unemployment, Salmon Point's project emphasized a curriculum featuring individualized instruction and offering vocational options both for graduates who planned to remain in the area and those who planned to leave Salmon Point (approximately 70% did each year). Such skills as first aid and how to complete an income-tax form were offered in addition to academic subjects. On Career Day, community members gave classroom lectures about their work. A "basic school" and a "conventional school" both offered innovative approaches to traditional curriculum. An extensive student-travel program addressed the problem of Salmon Point's extreme isolation. Preschool education and community education programs were also instituted.

PRAIRIE MILLS (SEVENTEEN COMPONENTS)

Faced with problems of a high outmigration rate of its young people, a large percentage of high-school dropouts, and a lack of student interest in continued education after high school, Prairie Mill's project included individualized instruction using packaged math and language arts programs, on-the-job training, early childhood education, career education, and adult education. A rugged obstacle course and primitive camping were part of the physical- and outdoor-education components.

BIG SKY (FIVE COMPONENTS)

Big Sky was one of the few school districts in the Rural ES program which did not emphasize basic skills. The focus of its ES project was on unifying its widely scattered, recently consolidated schools. To this end, the project offered an extensive selection of adult education courses and a variety of cultural events. The project also included career education, backed up by guidance counseling and a media center, which picked up satellite transmissions of television programs related to career education in addition to serving as a resource center for teachers.

CLAYVILLE (ELEVEN COMPONENTS)

New residents seeking employment in the burgeoning industrial development around Clayville in the 1960s brought with them new ideas about schooling which often conflicted with those of older residents and affected the school system as well. In the ES project, community–school committees fostered cooperation among community members and personnel in the schools; these committees arranged cultural events and collected the history of the region. The ES project also implemented individualized instruction through packaged programs in reading, math, and science to deal with a student population which had had a variety of previous educational experiences.

BUTTE–ANGELS CAMP (TWELVE COMPONENTS)

With the largest population of all the 10 rural districts, this project emphasized individualized instruction through packaged reading and math programs and through "diagnosis and presciption," in which each child's education was designed to suit his or her needs. Team teaching and the use of aides also had an impact on changing the teacher's role. An environmental-education component used the outdoors to teach science and social aspects of the environment. A new system of governance, in which community people were included in a committee to propose changes in the schools, was attempted, as well as similar groups involving teachers.

LIBERTY NOTCH (ELEVEN COMPONENTS)

Liberty Notch (three autonomous school districts which shared a common central office) suffered from a depressed economy. Its project emphasized career education, including on-the-job training for students in a local hospital and other work sites, and psychological services for students. A language arts program involved all grade levels in integrated, sequential curriculum. An environmental-education component included nature walks, camping trips, housing construction, and courses in geology, geography, and meteorology. An adult education component served the needs of a poorly educated population.

MAGNOLIA (EIGHT COMPONENTS)

Magnolia faced a high rate of unemployment, a poorly educated adult population, poor pupil performance on national achievement tests, and few services such as guidance counseling that could be used in conjunction with school programs. To meet these problems, Magnolia planned and implemented an early childhood education component and a career education component, health and physical education, and competency-based instruction, in which information was collected about each student's skills and attitudes in each subject area and appropriate learning activities were developed.

OYSTER COVE (FOURTEEN COMPONENTS)

Oyster Cove emphasized individualized instruction using several packaged reading programs, along with career education, environmental education, and "interim week," which took students on field trips during semester break. Arts and crafts, auto mechanics, and independent study (correspondence courses) were also a part of Oyster Cove's project. A community education component services the local community.

TIMBER RIVER (FORTY COMPONENTS)

The approach taken by Timber River to construct its local ES project was to amass a great number of discrete and highly differentiated components. Its 40 components are more than double the number included in the project of any other school district. Career education was one focus of the project, in part because of the district's high rate of seasonal unemployment in its major industry—logging. Other areas which were addressed included expanded staffing and varying the curriculum. Five components were designed to make schooling schedules more flexible.

DESERT BASIN (NINE COMPONENTS)

With 25% of its population of Mexican-American heritage, Desert Basin was a community divided into two segments with differing educational needs. Its ES project included a reading program based on diagnostic methods, early childhood education, bilingual instruction, counseling, and adult education. A media component in which students and community groups helped to run a cable-television station in the school became widely known around Desert Basin.

THE ASSESSMENT OF EFFECTIVE INITIATION

Although each of the 10 rural school districts eventually produced a formal project plan acceptable to the staff of the ES program, they differed greatly in the process used to produce their plans. We now consider factors which can help to explain some of this variation in the planning process.

Based upon our review of the available literature and an analysis of available data about these 10 school districts (these include the formal project plans, the ethnographic case studies, periodic interviews with each of the OSRs, and questionnaires completed by professional personnel in each district), we have identified five factors which may be used to set criteria to judge the effectiveness of initation of the ES projects.

1. *Breadth of participation in determination of needs.* It is generally maintained that the wider the participation of diverse groups in the identification of needs for an innovative project, the lower the probability that serious needs will be overlooked (Shepard, 1967). One of the major expectations of ES/Washington during the planning year was that the 10 districts would reexamine their needs through broad-based participation.[4]

In assessing the effectivness of initation for each school district, we have assumed that those districts in which there was a broad participation by potentially affected groups in the determination of needs (through meetings and surveys) would be better prepared to implement their projects than those in which there was more limited participation.

2. *Degree of broad-based influence on formal project plan.* It is widely assumed in the literature on planned change that a decentralized decision-making structure during the planning period of an innovative project will facilitate subsequent implementation (Coughlan, Cooke, & Safer, 1972; Zaltman *et al.*, 1973). Although this assumption was also held by the staff of the ES program, and frequent efforts were made by them to facilitate such participation, there was a marked tendency within these 10 school districts for decision-making power to be located in the office of the district superintendent, with occasional input from building principals. District school boards had noticeably less influence, followed by teachers, parents, and other citizens. However, there was variation among the ten districts in the degree to which decisions about the formal project plan were the result of a broadly based process.

In order to ascertain the breadth of influence in the formal project plan, we asked all professional staff members in each school district to estimate the degree of influence of five actor groups: superintendent, principals, teachers, parents, and other community groups.

[4]Breadth of participation was assessed as follows, based on data derived from interviews with OSRs:

- A *high* degree of participation was assumed if information was gathered from groups both within the school system and the wider community.
- A *moderate* degree of participation was assumed if information was gathered from several groups within the school system, including teachers as well as administrators.
- A *low* degree of participation was assumed if information was gathered primarily from the administrative leadership of the school system.

In assessing the effectiveness of initation for each school district, we have assumed that those in which there was broad-based influenced by these potentially affected groups on the formal project plan would be better prepared to implement those plans than those in which there was a more limited influence.[5]

3. *Degree of congruence between perceived problems and project goals.* There is no guarantee that all information gathered through such a process of "participatory democracy" as that advocated by the ES program will have an equal opportunity to be incorporated into whatever decisions are subsequently made about project priorities. Cooke (1972) points out that within complex organizations, collective decision making must operate in coordination with formal authority structures. Decisions made by advisory bodies are geneally subject to review and revisions by the organization's hierarchy of authority, and in such cases information is often distilled, distorted, or ignored.

It is quite likely that during such a process of review and revision within each of these 10 school districts, people in position of formal power interpreted project goals in a way that neglected the preferences of some groups who would be affected by the project. One need not postulate undemocratic tendencies on the part of those in positions of power to make such an assumption. In the hustle and bustle of bringing together diverse sets of opinions, and in organizing them into a planning document acceptable to an outside agency, many opinions and preferences can be overlooked. Nevertheless, Greenwood *et al.* (1975) report that innovative projects which were judged to be solutions to widely acknowledged local problems were more likely to be successfully implemented than those which were not.

We have assumed that school districts in which there was a high degree of congruence between perceived problems (as reported by educators and community residents) and project goals (as stated in the

[5]Degree of broad-based influence on the formal project plan was assessed as follows, based on data derived from responses of professionals on a survey administered in the fall of 1973:

- A *high* degree of broad influence on the formal project plan is assumed if *at least four* of the five groups were viewed by at least 50% of the professional staff to have had at least "some influence" in the planning process. (See Table 6.)
- A *moderate* degree of broad influence on the formal project plan is assumed if *only three* of the five groups were viewed by at least 50% of the professional staff to have had at least "some influence" in the planning process.
- A *low* degree of broad influence on the formal project plan is assumed if *only one or two* of the five groups were viewed by at least 50% of the professional staff to have had at least "some influence" in the planning process.

formal project plan) would be better prepared to implement their projects than those in which there was a lesser degree of such congruence.[6]

4. *Independence from the external funding agency in formulating the formal project plan.* As noted earlier in this chapter, ES/Washington anticipated playing an active role in the monitoring of the grant relationship between the federal government and the 10 school districts. These expectations seemed to stem from a strong desire on the part of the staff of the ES program to be of help in what was generally agreed to be a most ambitious undertaking on the part of these school districts.

Public statements made by the U.S. Commissioner of Education while meeting with representatives from six of the 10 school districts in Washington, D.C. in July 1972 emphasized a federal role of "helping local educational agencies to help themselves." Such a message was also conveyed by the ES/Washington program officer for these sites during visits in August 1972. However, as all 10 districts began to experience difficulty in understanding the elusive concept of "comprehensive change," in eliciting broad-based participation in the determination of goals, in linking project procedures to goals in a rational way, and in rationally assigning anticipated federal dollars to project procedures, there developed many instances in which the federal role was viewed as one of "interference" rather than of help. In some cases, perceptions of such interference had only minor impacts upon success in subsequent stages of the change process, but in most instances such concerns became a major obstacle.

Because of the preponderant equation of federal "influence" with "interference," we have assumed, in assessing the effectiveness of initiation for each school district, that those in which the ES/Washington staff were viewed by the local school district staff to have had only

[6]In order to ascertain the extent to which each formal project plan reflected widely acknowledged local problems, we elicited the questionnaire responses of two groups (all professional staff of the school system, and a random sample of community residents) about local problems. We then compared their statements of major local problems with the goals stated in the formal project plan for that school district. Based on the judgment of members of the research staff, we grouped the districts as follows:

- A *high* degree of congruence was assumed if the formal project plan directly addressed major problems reported by the professional educators.
- A *moderate* degree of congruence was assumed if the plan indirectly addressed major problems reported by professional educators and community residents.
- A *low* degree of congruence was assumed if the plan did not address major problems reported by professional educators and community residents.

minor influence upon the formal project plan would be better prepared to implement their projects than those in which they perceived a greater degree of influence.[7]

5. *Degree of acceptance of the project by the professional staff.* In Chapter 4, we emphasized the importance of the commitment of the administrative leadership of a school system to the achievement of a readiness to embark upon a process of externally induced planned change. We argued that unless key administrators were directly involved in the negotiations with the external agency, the school system could be at a disadvantage when attempting to initiate the change process. Although the importance of the involvement and commitment of key administrators remains high during the initiation stage, successful completion of this stage seems also to require the involvement and commitment of other members of the organization who will be affected by the planned change, particularly the instructional staff. It is widely acknowledged in the research literature that if teachers are left out of the planning process or if they feel that their participation was perfunctory, they are likely to view any formal project plans as unrealistic and unworkable and to become serious obstacles to successful implementation.

In assessing the likely effectiveness of initation for each school district, we have assumed that those districts in which the teachers viewed their ES project favorably would be better prepared to implement their projects than those in which the teachers had less favorable views.[8]

[7]Data were derived from local school-district staff response on the fall 1973 questionnaire. We grouped the districts according to the following criteria:

- A *high* degree of independence was assumed if less than 30% of the district professional staff reported that ES/Washington had "a great of influence" on the formal project plan.
- A *moderate* degree of independence was assumed if between 30% and 60% of the district staff reported that ES/Washington had "a great deal of influence" on the plan.
- A *low* degree of independence was assumed if more than 60% of the district professional staff reported that ES/Washington had "a great deal of influence" on the plan.

[8]This factor was assessed on the basis of questionnaire responses in the fall of 1973.

- A *high* degree of acceptance was assumed if more than 80% of teachers were reported to view their district's participation in ES as "a good idea."
- A *moderate* degree of acceptance was assumed if between 55% and 80% of the teachers were reported to view their district's participation as "a good idea."
- A *low* degree of acceptance was assumed if less than 55% of the teachers were reported to view their participation in ES as "a good idea."

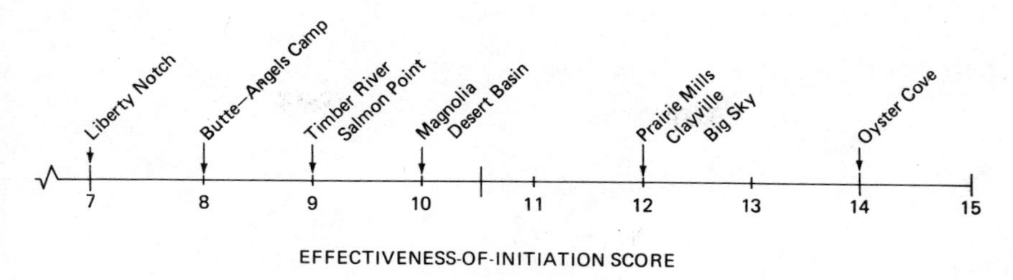

Figure 3. Effectiveness-of-initiation scores for the 10 school districts.

A Summary of Initiation Factors

The diversity of the 10 school districts in the effectiveness-of-initiation factors is displayed in Table 5. No two of the 10 districts presented the same "profile" on these factors. Their profiles range from that of Oyster Cove (high on four factors and moderate on one) to Liberty Notch (low on three factors and moderate on two).

Following a procedure similar to that used to compute a "best estimate" of readiness for each school district (see Chapter 4), we have used the factors discussed in this chapter to compute an "effectiveness-of-initiation" score for each district.

In order to compute this score, we have again made some simplifying assumptions. We have assigned the numerical scores of 3, 2, and 1 to the categories of "high," "moderate," and "low," respectively, for each factor of initiation.[9] We then computed an overall "effectiveness-of-initiation" score for each district by adding together its scores on each of the factors. Thus, Oyster Cove is estimated to have had the highest "effectiveness-of-initiation" score (see Figure 3). It obtained a score of 14 by receiving 3 units for being high on four of the factors and 2 units for being moderate on staff acceptance of the project. Liberty Notch is estimated to have had the lowest "effectiveness of initiation" (7 units). It obtained this score by being moderate on breadth of participation in determining needs and breadth of influence on breadth of participation in determining needs and breadth of influ-

[9]Each district was ranked independently by three members of the project research staff. Consensus was then reached by rating the districts in three categories—high, medium, and low—for each factor. It is important to note that our approach makes a series of "simplifying assumptions." We recognize that other plausible assumptions could be made which might lead to a different classification of the 10 districts in terms of their overall effectiveness of initiation.

Table 5. Effectiveness-of-Initiation Profiles for the 10 School Districts

School district[a]	Effectiveness-of-initiation factors				
	Breadth of participation in determining needs	Breadth of influence on formal plan	Goal–problem congruence	Independence from NIE influence	Staff acceptance of project
Oyster Cover	H	H	H	H	M
Prairie Mills	M	M	H	M	H
Clayville	H	L	H	M	H
Big Sky	H	H	M	M	M
Magnolia	M	H	M	L	M
Desert Basin	H	M	M	M	L
Timber River	M	L	H	M	L
Salmon Point	L	M	L	H	M
Butte–Angels Camp	M	M	L	M	L
Liberty Notch	M	M	L	L	L

[a]The order of school districts corresponds to their overall effectiveness-of-initiation score. *Key:* H = high, M = moderate, L = low.

ence on formal plan, and low on the three other factors. (The predictive utility of this scale is considered in Chapter 9.)

WHOSE PLAN?

A primary focus of the factors of initiation is on ownership of the final plan; given the original assumptions and the expectations of the ES program, this is a crucial issue. After the extensive and intensive interactions of federal and local officials during the planning process, one might ask whose plan it was that served as the basis of the formal contract signed with NIE for implementation of "local" projects.

The five factors of initiation may be seen as criteria for assessing the *process* of initiation. However, the *outcomes* of the process—the plans themselves—had to meet additional criteria in order to be acceptable to ES/Washington. These included a central theme which was comprehensive, attention to all the facets of changes, and clearly stated goals and objectives and rational *methods for achieving them* which would serve as a blueprint for implementation.

A comparison of the two sets of criteria makes it clear, however, that a district could carry out a highly effective process of initiation and produce a plan unacceptable to ES/Washington or, conversely, produce a plan acceptable to ES/Washington by means which did not represent an effective initiation process. Big Sky is an example of the former. Its process of initiation ranked high, but the resultant plan could not meet ES/Washington's criteria for comprehensiveness and (as mentioned earlier) eventually had a perfunctory central theme imposed on it. In contrast, local planning personnel in Desert Basin produced an acceptable project plan in close cooperation with ES/Washington. However, Desert Basin's process of initiation was flawed; staff acceptance was too low for the plan to survive the departure of its planners. The outcome in both districts was negative in its impact on the future of the local projects.

The partnership of the federal and local agencies in the ES program seems to have depended on a congruence between the expectations ES/Washington had for the program and the needs the districts were expected to perceive during the planning process for their local projects. The testimony of the planning year is that for many of the districts this congruence did not obtain, and that the compelling need of the districts to produce an acceptable plan to receive their funding changed the equal partnership during the planning years into a "senior-junior" partner relationship, with ES/Washington as the senior partner.

The final "catch" in the federal-local relationship was ES/ Washington's expectation that the formal project plans would serve as blueprints for local implementation. However, as we have seen, the priority of the districts was generally on production of the plan as an end in itself—that is, to acquire badly needed long-term funding. Although pilot implementation had been encouraged during the planning year, residual problems associated with the planning process had to be faced in September of 1973, when large-scale implementation was to begin. There were two major sources for these problems. First, the planning, which had been done primarily *at the district level,* had to give way to implementation *at the school level.* Second, the task of making the plan acceptable to ES/Washington had in some cases caused the districts to lose sight of their early ideas concerning local needs, and had resulted in a relationship between ES/Washington and core local administrators which left teachers and community disaffected with the final project plan.

The demands of the planning year, particularly those imposed by ES/Washington, tended to narrow the planning process at the local level progressively toward a small core group of administrators, who dealt with ES/Washington on the matters of plan revision, and being under time constraints, consulted less and less frequently with other constituencies—particularly teachers and other citizens.

In Prairie Mills teachers had put in time without pay to develop project components, yet in the course of rewriting the project plan three times, many of their ideas were abandoned or modified. It was the teachers' perception that this had taken place because some "administrator had a different idea," and many of them were aggrieved (Donnelly, 1979b, p. 200). Similarly, when teachers in Butte–Angels Camp read the final project plan it appeared to them that although their ideas had been solicited early in the planning process, the writers of the final plan had willfully eliminated those ideas and that the real influence of the ideas had been minimal (Firestone, 1979, pp. 167–168). Perhaps the most extreme case was that of Clayville. There the superintendent had attempted to impose the ES project on the community and the school district from the outset. When the implementation effort began he addressed the teaching staff and announced, "This is a blueprint and this is going to be the way we operate the schools" (Clinton, 1979b, p. 227).

These examples are not exceptional. According to a survey conducted in the fall of 1973, teachers in every district except Oyster Cove felt that they had had less influence over the planning process than either the superintendent or the principals (Table 6). For that matter, respondents saw the principals' influence as being less than that of the superintendent in all districts. All groups perceived the superintendent as extremely influential. Thus, the ES planning process was generally perceived as a "top-down" phenomenon; when district-wide planning was complete and it came time to "move down" to the school level, teacher attitudes would very likely weigh heavily on the success of implementation.

Districts tried various strategies during the move down to the school level; the most important was the use of coordinators of various kinds. Three different methods of using coordinators illustrate the spectrum of ways in which districts attempted to bridge the gap between planning and implementation.

Timber River created positions for five "instructional leaders"—one for each school building—and a district-wide coordinator of staff development. Five of these six were teachers recruited from within the system. The teachers were generally enthusiastic, since the instructional leaders were not only planning and coordinating activities but were being "consistently helpful"; they even helped out in the classroom. Some problems developed, especially with principals, because of role ambiguity: Were the instructional leaders responsible to the principals, the coordinator of staff development, or someone else in the central office? Nevertheless, the experience was on the whole positive, and although the positions were scheduled to end with the end of ES funding, there was considerable support for their continuation at local expense (Hennigh, 1979).

Coordinators meant something else entirely in Desert Basin. When the superintendent and the core group responsible for planning left the district in the spring of 1973, the new superintendent filled all but one of the six coordinator positions with outsiders. This group became known as the superintendent and his "cadre of experts." The superintendent was perceived locally as an ambitious person who would go on to a bigger school system and who was using Desert Basin

Table 6. Percentage of Professional Staff Members Reporting that Various Potentially Affected Groups Had at Least "Some Influence" on Their District's Formal Project Plan, by School District[a]

School district	Potentially affected group				
	Superintendent	Principals	Teachers	Parents	Community groups
Salmon Point	100%	85%	62%	42%	33%
Prairie Mills	100	96	82	45	26
Big Sky	84	80	47	62	63
Clayville	97	59	40	41	46
Butte–Angels Camp	99	68	58	36	34
Liberty Notch	99	68	58	36	34
Magnolia	98	86	59	62	48
Oyster Cove	80	70	90	80	67
Timber River	100	59	36	21	22
Desert Basin	92	77	26	41	54

[a]Data are based upon the responses of professionals in the fall of 1973 to the following questionnaire item: How much influence did each of the following groups have in determining the multiyear plan submitted by your school district to the Experimental Schools program in Washington, D.C. in the spring of 1973? The response alternatives were: "A great deal," "Some," "Very little," "Almost none," "None," and "I don't know."

as a stepping stone. (He did indeed depart for a larger school district at the end of the first implementation year.) He was perceived to have brought in a staff whose primary loyalty was to him personally; thus he and his experts would assume ownership of the project. The subsequent story of implementation in Desert Basin was one of the isolation of the project from the district as a whole and of the coordinators from the overall administrative structure of the district (A. Burns, 1979).

Liberty Notch faced still another kind of implementation problem. An ES project director and three coordinators were in charge of the project. No clear-cut course of action was laid out in the formal project plan. Furthermore, the project director and the three coordinators had to report to three separate school boards. The jealously guarded autonomy of the three separate districts did not make their task easier.

The move from planning to implementation was most painful in Liberty Notch, but in virtually all districts the beginning of the implementation year was a period of overwhelming transition. The assumptions of the ES program and the expectations for the planning year were based on the presumption of tightly linked federal-local and school district–school relationships which did not exist. The consequences of the early misunderstandings and disappointments were never fully overcome. As will be seen in the following chapters, implementation of the ES projects was uneven, both across districts and among schools within districts. However, despite all of the problems, changes were implemented. Over time, the federal-local linkage became looser and the school district–school linkages tighter. Local resentment and bitterness faded. As teachers became aware that elements of the ES project were not totally foreign to their own priorities, many went about the business of implementing elements of the projects in their schools (Stannard, 1979).

With this background of planning and implementation at the district level, we now turn to the implementation of changes in the schools, the focus of the next two chapters.

6

The Implementation of Planned Change in Schools

The Rural ES program was designed to be a district-wide effort encompassing change in five distinct facets of school-district activity. However, many of the components of the local ES projects were tailored to individual schools or grade levels within each district. This was particularly true of changes in curriculum, with elementary-school programs often directed at basic skills (such as reading) whereas high-school programs often emphasized specialized subjects (such as career education). Since over 70% of the components implemented by the 10 districts were in the area of curriculum and therefore occurred within schools or classrooms, we must examine implementation at the school level, although the support and management of change from the central office in the school implementation process was undoubtedly important as well.

Even those components which were designed to be implemented throughout a district were often implemented to different degrees and in different ways in its schools. Given the "structural looseness" which often characterizes school districts (Bidwell, 1965; Deal *et al.*, 1975; Weick, 1976) and the degree of autonomy maintained by individual schools, an examination of differences in implementation among schools is crucial to understanding the process of change within school systems.[1]

The objectives of this chapter are to examine the differences in implementation among schools and to begin to answer the central question: Why was implementation greater in some schools than others? To achieve these objectives we first introduce the dimensions by which we have measured the scope of implementation at the school level. Second, we provide a description of the degree to which implementation actually occurred in the schools and the variation in degree of implementation among schools. Third, we introduce the variables we used to define the system characteristics used in our analysis. Finally, we present an analysis of how system characteristics appear to predict implementation at the school level and relate our findings both to our theoretical framework and to previous research and theory. But before turning to the matter of implementation, we briefly describe the schools which participated in the ES program.

The 49 schools[2] vary considerably in grade spans, dispersion within school districts, and size. Most (28) of the schools are elementary schools, eight of which cover grades K–8. Of the remaining 21 schools, five are middle or junior high schools, four are combined junior-senior high schools, 11 are senior high schools, and one school covers all grades K–12. Since most of the school districts in this study have been created through consolidation efforts, some of the schools are located long distances from the central district office. Although 11 of the schools are located next door to or in the same building as the

[1]An additional practical reason for examining change at the school level emerges from a desire to maximize the degree to which we are able to examine multivariate relationships between variables which may be related to effective implementation of change and the scope of change. When the school district is the analytic unit, considerable limitations are placed upon our analysis, since only 10 school districts participated in the Rural ES program. When the analytic unit is schools, on the other hand, a larger population (the 45 schools for which there were sufficient data out of the 52 participating schools in the 10 districts) allows us to examine the relationship between several variables and scope of implementation simultaneously, and consequently to extend the reach of the analysis.

[2]The maximum number of schools in the ES districts was orginially 52, but three of the elementary schools are totally excluded in the analysis. One of them is a "ranch school" having one teacher and two pupils. One was dropped from the study because it is located in an adjacent school district which sends pupils to an ES district high school, and later dropped out of the ES project. One was a small "annex" school buidling and was not used each year. However, due to the smallness of some of the schools and their limited response rates on surveys, most of the analyses presented in this volume are based on *N* of 45 schools.

central office, the remainder are an eighth of a mile or more away. Indeed, seven of the schools are more than 20 miles away from the superintendent's office, the most distant one being 64 miles away. The number of schools within each district ranges from two to 12. The distributions of school size, both in number of teachers and number of pupils, are presented in Table 7.

MEASURES OF IMPLEMENTATION AT THE SCHOOL LEVEL

The measures that were used to describe implementation at the school level were derived by using a modification of the scope-of-implementation matrix presented in Chapter 3. The objective was to obtain school-based scores for the two dimensions of organizational change—quantity and quality—which together would constitute a total "scope-of-implementation" score. The data were acquired from forms that were designed to provide information about implementation of ES activities in each school. These school-oriented forms were completed by our OSRs with the aid of local informants. The brief descriptions below list the variables that compose the three measures of implementation.

Quantity

Quantity of change is the dimension which measures the degree to which innovations have penetrated the system. Quantity was calcu-

Table 7. Characteristics of Schools in the ES School Districts (N = 49 Schools)

		Range	
Characteristics	Mean	Lowest	Highest
Number of schools in district	5	2	11
Number of teachers			
All schools	18	1	44
All elementary	11	1	27
All secondary	20	7	44
Number of pupils			
All schools	284	13	739
All elementary	235	13	587
All secondary	355	73	739

lated by adding four variables: the percentage of students in the school who participated in ES programs, the percentage of teachers who participated, the average involvement of students (expressed as a percentage of the school day) who participated, and the average involvement of participating teachers.[3] Subdimension scores of pervasiveness and extent described in Chapter 3 were also calculated. Pervasiveness was measured by adding the percentages of teachers and students involved, while extent was measured by adding the average percentages of involvement of affected teachers and students. These two subdimension scores, as well as the quantity score, were standardized to range from 0 to 100.

Quality

The quality dimension reflects the "degree of difference" from what existed before the changes. Quality is measured by adding five variables which reflect the OSR's judgments about changes in (1) the use of time, space, and facilities; (2) the level of community involvement; (3) the administration and governance of the school; (4) the curriculum; and (5) the school structure.[4] The quality score was standardized to range from 0 to 100.

[3]In computing quantity, teachers and students were treated as two equal groups. The resultant quantity score thus gives greater weight to teachers, although it does take into account variation in student–teacher ratios in schools. Descriptive statistics for the four quantity variables follow ($N = 49$ schools):

Variable	*Mean*	*Median*	*S.D.*	*Minimum score*	*Maximum score*
Students affected, %	77.5	99.8	.32	8	100
Teachers affected, %	75.5	95.0	.31	0	100
Student time affected (only for those involved in ES), %	20.9	12.3	26	10	90
Teacher time affected (only for those involved in ES), %	32.2	30.0	.26	0	90

Cronbach's Alpha coefficients of internal reliability for these four variables was .60.

[4]Changes in the use of *time, space, and facilities* were computed by averaging separate indicators for each of the three parts of the facet (time, space, and facilities). (OSR Questionnaire, Part II, Item 9 in Appendix B.)

Change in *community involvement* were measured by averaging two indicators, one of which asked the OSR to indicate the level of change in involvement of community members as *active* participants (as students or as supplementary instructors) in educa-

Total Scope Score

The total school scope was computed by adding together the dimensions of quality and quantity. To verify that the nine variables included in total scope were measuring the same construct, we computed the internal reliability of the measure. Cronbach's Alpha coefficient of internal reliability was .76.

---

tional activities, and the other about changes in the level of involvement in policy meetings regarding school activities. (OSR Questionnaire, Part II, Items 11 and 12 in Appendix B.)

Changes in *administration and governance* were measured by taking the average of two indicators. The first of these represented the mean of change in the influence of students, parents, teachers, principals, the superintendent, and the school board as a result of ES. (OSR Questionnaire, Part II, Item 10.) The second represented an average score for two items asking about changes in lines of authority and decision making. (OSR Questionnaire, Part II, Items 14a and b.)

The three variables above were each measured on a scale whose minimum was 0, and whose maximum was 5. The mean and standard deviation for each of the variables are as follows ($N = 49$ schools):

Variable	*Mean*	*S.D.*
Time, space, and facilities	1.68	.97
Community involvement	1.13	.90
Administration and organization	1.17	.80

Changes in *curriculum* and changes in *school structure* were each measured by a single indicator with ordinal categories. (OSR Questionnaire, Part II, Items 16 and 17.) The former indicator asked the OSR to rate the degree to which ES funded changes in curriculum represented major vs. minor changes. The latter asked whether participation in the ES program has resulted in changes in the organizational structure of the school at either the classroom or the entire school level. Distributions of responses for these categories are shown below ($N = 49$ schools):

Variable/response category	*Percentage distribution*
Curriculum	
No change	12
Shift in emphasis	19
Addition of new elements	61
Substitution of new elements for previous curricula	9
	100
School structure	
No change	39
Change primarily at classroom level	41
Changes affecting entire school; replace previous structures	20
	100

A computational problem in creating a measure for quality was the fact that three of the scales ranged from 0 to 5, one was measured on a 0–3 scale, and the fifth on a 0–2 scale. Since we wished to allow all indicators the opportunity to contribute equally to the final score, the two shorter scales were simply recorded to range from 0 to 5. Cronbach's Alpha coefficient of internal reliability for these five variables was .76.

To facilitate comparison with dimension and subdimension scores, total scope was also standardized to range from 0 to 100.

A DESCRIPTIVE ANALYSIS OF THE SCOPE OF IMPLEMENTATION IN SCHOOLS

Table 8 presents some descriptive statistical results of the effort to measure changes in quantity and quality occurring in the schools during implementation of the ES projects. Several conclusions may be drawn about implementation of the ES projects from the results of these simple measurements. First, it becomes apparent that the Rural ES program was achieving at least one of its mandates, to involve students and staff members in new programs: the mean school score on the quantity dimension was 54 out of a possible 100. The statistics for the two subdimensions of quantity show that pervasiveness was high; the mean percentage of the number of people involved in the Rural ES program across all the 49 schools was 77. Furthermore, over half of the schools had pervasivness scores of 83 or higher. The extent score was relatively low; students and staff affected by an ES project spent an average of 31% of their school day on ES related activities. Although this score falls far short of the theoretical limit of 100%, it hardly represents a failure of the Rural ES program to implement comprehensive change. Rather, given the documentation about the difficulty of instituting widespread changes within schools (see, e.g., Sarason, 1971), the fact that in the aggregate the Rural ES program affected over one-fourth of the day of three-fourths of the teachers and students within these 49 schools suggests that substantial involvement in the implementation effort was taking place.

However, neither pervasiveness nor extent of overall implementation adequately reflects the degree to which the new activities were qualitatively different from those which existed before participation in the Rural ES program. When we look at the dimension of quality, we find that the typical school had a score in the middle thirties, indicating an achievement of only 30% of the possible score which could be reflected in the measures of our five quality variables.[5]

[5]Furthermore, the standard deviation on the quality dimension is also lower than for the quantity dimension. This is largely due to the fact that no school scored over 57 on this scale, and the observed range of scores is, therefore, depressed.

Table 8. Descriptive Statistics for Scope of Implementation in Schools by Dimension (N = 49)[a]

Dimension	Mean	Median	S.D.	Lowest	Highest
Quantity	53.6	55.0	20.4	15.5	95.0
(Pervasiveness)	(76.6)	(83.3)	(25.4)	(8.3)	(100.0)
(Extent)	(30.6)	(26.8)	(25.0)	(10.0)	(90.0)
Quality	32.6	35.7	14.1	2.5	57.0
Total scope	43.1	39.5	15.5	13.5	76.0

[a]The subdimensions of quantity and their descriptive statistics are presented in parentheses.

Overall, the schools in the 10 districts in the Rural ES program failed to change as much as was theoretically possible. The average school achieved a total scope-of-implementation score of 43, or less than 50% of the possible score (Table 8). However, the failure of the schools as a group to change radically does not necessarily mean that the program was a failure; important, though not dramatic, changes were occurring in many of the schools. The fact that 17 out of the 49 schools achieved at least 50% of the possible maximum scope-of-implementation score, and that four schools achieved 70% or more of the maximum scope score (see Table 9), represents in great measure an achievement for the Rural ES program as a whole. In addition, each of the Rural ES sites had at least one school that scored over 40 out of a

Table 9. Percentage and Number of Schools Falling into Each Tenth of the Possible Range of Total School Scope Score (N = 49)

Tenth of possible range	Percentage of schools	Number of schools
10th (highest)	—	—
9th	—	—
8th	8	4
7th	4	2
6th	23	11
5th	14	7
4th	35	17
3rd	10	5
2nd	6	3
1st (lowest)	—	—
	100	49

Mean total scope score = 43.1; standard deviation of observed scope scores = 15.5; minimum observed scope score = 13.5; maximum observed scope score = 76.0; lowest possible total scope score = 0.0; highest possible total scope score = 100.0.

possible 100 on our scale. Thus, even within a district which generally "failed" to introduce major comprehensive changes throughout the system, at least one of its schools stands out as being innovative by our criteria. Although the Rural ES program does not appear to have fully achieved its objectives, the incremental changes associated with ES funded activities in individual schools point to effects that should not be ignored or subsumed under blanket statements about the resistance of schools to change.

VARIATIONS IN IMPLEMENTATION WITHIN DISTRICTS

At the outset of this chapter we suggested that the school is the most reasonable level at which to perform most of our analyses of implementation. If all schools within each district implemented changes equally, we could conclude that characteristics in the environment of the district were overwhelming any propensity for schools to behave autonomously. However, schools in many of the districts varied considerably in their scope of implementation of the ES project (Figure 4). Six of the 10 districts exhibit wide ranges (more than a

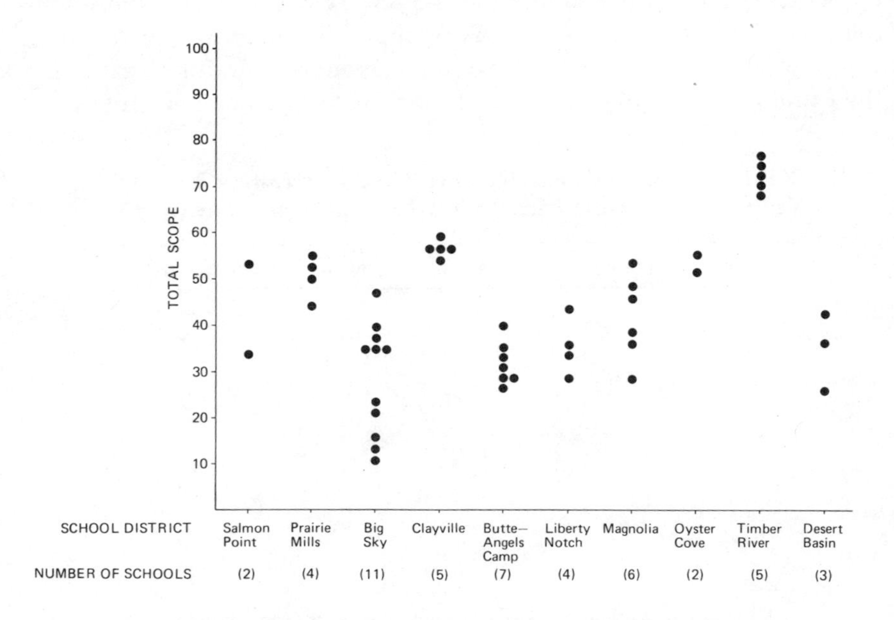

Figure 4. Total scope score for schools by school district ($N = 49$).

10-point difference between the lowest and highest scoring school) in the overall scope-of-implementation scores achieved by their schools. In Salmon Point, for example, one of the two schools exhibits a relatively high implementation score, whereas the other school scores quite low. In Big Sky, with 11 schools, the scores range from approximately 15 to 45. In the four districts which show more of a clustering of schools in their scope scores (Prairie Mills, Clayville, Oyster Cove, Timber River), there is still some variation.[6]

The descriptive investigation of variability in implementation is only the first step in answering the more critical question: How can we explain the variability in implementation and why was the Rural ES program more successful in some schools than in others?

SOCIAL SYSTEM CHARACTERISTICS: THE VARIABLES AND THEIR MEASURES

As noted in Chapter 2, it was our hypothesis that variations in the social system would go far in explaining variation in implementation among schools. To test this hypothesis, we first had to select variables representing each system element. In particular, we looked at the *structure* of the schools, in which we included variables defining the formal properties of the organization and its decision-making system; the *culture* of the schools, including variables defining the informal values and norms that pervade the organization (these variables are sometimes defined as organizational "climate," or the internal environment of the organization); and the *input* to schools, which was defined for the purpose of this analysis as comprising school-staff characteristics.

Structure and culture variables were measured in the fall of 1973, at the end of the project's planning year. Teachers in the districts were mailed a questionnaire which covered a wide variety of issues regarding the operations of their school and districts. Since the purpose of this study was to look at the behavior of schools as organizations rather than at the behavior of teachers, teacher responses within school were

[6]Some of the reasons for differences in interschool variation among the districts will be discussed in Chapter 8.

averaged to obtain a school score for each variable. Thus the variables discussed below represent structure and culture as perceived by those teachers responding in each school.

The fall 1973 questionnaire was anonymous and included no data on the personal characteristics of the staff. However, data from other questionnaires administered to the professional personnel of the 10 ES districts were collected in the spring of 1973 and in the spring of 1974, and are the source for our input variables. (However, only data on staff members who were employed in the fall of 1973 are used.) These data were also aggregated to the school level, and average scores or percentages distributions were computed based on the number of respondents per school.

Structure Variables

The structure variables were selected on the basis of a growing consensus concerning the crucial dimensions of formal organizations (see, e.g., Blau & Schoenherr, 1971; Hage & Aiken, 1970; Pugh *et al.*, 1968). The structure variables include *complexity, size, formalization, individualized technology, classroom autonomy,* and four variables that deal with the *structuring of authority* in the school. Anther structural variable unique to schools is that of *school level* (elementary vs. secondary). With the exception of size and level, each of these variables was scaled from multiple items on the questionnaire. (The variables, their operational definitions, and Cronbach's Alpha coefficient of internal reliability are presented in Table 10. Each variable met the minimum criterion we had established prior to conducting the analysis—Cronbach's Alpha of at least .65.)

Culture Variables

Although organizational researchers generally agree that the informal properties of an organization will affect the change process in at least some ways, most research on the impact of the informal organization on change has been in the form of qualitative case studies rather than quantitative studies using large samples of organizations. As a

Table 10. School Structure Variables: Operational Definitions and Coefficients of Internal Reliability

Variable	Operational definition/measures used	Data source	Cronbach's Alpha coefficient of internal reliability
Complexity	No. of specialists in the school No. of administrators in the school No. of different occupational specialties in the school	1973 data provided by on-site researcher	[a]
School size	No. of pupils		[b]
School level	Dummy variable representing either elementary or secondary school (secondary = 1)		(not applicable)
Formalization	No. of formal policies that are regularly enforced	PPCF 1973,[c] Q23	.77
Individualized technology	Use of individualized instruction	PPCF 1973, Q22	.82
Classroom autonomy	No. of classroom decisions that the teacher can make on his or her own	PPCF 1973, Q21	.72
School board authority	No. of decision areas in which each group was felt to have influence (maximum = 13)	PPCF 1973, Q24	.82
Superintendent authority			.85
Principal authority			.84
Teacher authority			.75

[a]The complexity measures were so highly interrelated (minimum correlation of .75) that a single index was computed by standardizing and adding the three separate scores. No reliability coefficient was computed because of the small number of items in the scale obtained from lists of school employee names and their formal titles which were collected by the OSR.
[b]The number of full time equivalent (FTE) professional employees in the school was also measured, but was so highly correlated with the number of pupils (over .90) that the single-pupil measure was used. No reliability coefficient was computed for this single-measure item.
[c]PPCF = Professional Personnel Census Form; Q = question.

result, the literature offers very limited guidance for variable definition.[7]

In general, the literature is in agreement that two sets of culture or climate variables are extremely important in determining organizational process. These are the *morale* of the staff and its *cohesiveness* as a work group. Since there are "work groups" as such in schools, we have redefined the latter variables as the *level of collegiality* among the staff. However, other variables also seemed important in defining the normative character of the schools. The list that we finally developed represents a potpourri of factors found to be important in schools. These were selected on the basis of available theoretical discussions, case studies, and quantitative studies, not all of which were directly concerned with change. Among these are the *level of tension* between various groups, the *frequency of disputes* that occurred over school-related issues, the *orientation toward change* of the staff as a whole, the *orientation to pupil autonomy* in the educational process, the staff's *perception of problems* within their school, the staff's *goal differentiation,* and the degree to which the staff perceived a *discrepancy between their goals and the achievement of goals.* All of these items were scaled from multiple items on the questionnaire. (The variables, their operational definitions, and Cronbach's Alpha coefficient of internal reliability are presented in Table 11.)

Input (Staff Characteristics) Variables

Although studies of educational outcomes have had little success in finding correlations between staff characteristics and student achievement, the characteristics of individuals have often been found to be associated with innovative behavior (see, e.g., Rogers & Shoemaker, 1971). Few studies, however, have attempted to examine

[7]Some exceptions to this generalization should be noted. Bowers's (1973) study of the organizational-change techniques in a number of organizations used the Survey of Organizations instrument developed at the Institute for Social Research. The survey is rich in variables measuring organizational climate. However, the fact that the data were obtained from manufacturing organizations rather than schools limits the utility of the results for educators and educational researchers (Bowers, 1973, pp. 21–43). The Rand Change Agent Study (Berman & McLaughlin, 1975) included a number of variables subsumed under the general heading of "Organizational Climate" which correspond to our notions of culture, but do not reflect all of the factors that we have chosen to measure.

Table 11. School Culture Variables: Operational Definitions and Coefficients of Internal Reliability

Variable	Operational definition/measures used	Data source	Cronbach's Alpha coefficient of internal reliability
Level of tension	No. of role pairs that have at least "some" tension	PPCF, 1973 Q26	.82
Frequency of disputes	No. of issues that cause frequent disputes between various groups	27	.87
Morale	Discrepancy between actual and desired levels of personal influence on 13 items	25	.86
Change orientation	Additive score on 6 change attitude items	12 B,E,F,H,I,L 12 C,G,J,K,N,P,V	.71
Pupil autonomy orientation	Additive score on 7 pupil autonomy attitude items	21, K-P	.77
Collegiality	Additive score on 6 collegiality items		.69
Perception of problems	No. of areas perceived as moderate or serious problems in the school	9	.71
Goal differentiation	No. of goals considered to be "very important"	7	.74
Goal discrepancy	The sum over 12 goals of the difference between the importance of goals and how well they are being accomplished	7	.77

staff characteristics as organizational properties or to examine the relationship of aggregate staff characteristics to organizational innovation.

In considering the staff characteristics that are likely to be associated with organizational innovation, two different categories of variables should be included. First, there are variables that reflect the professional experience and behavior of the staff members. Within this group are the number of years of classroom *teaching experience,* the extent of staff members' *professional reading,* and the degree to which their experience as educators has been in contexts that are technologically advanced and cosmopolitan—that is, the "*modernity* index" developed by Herriott and Hodgkins (1973).

In addition, personal chacteristics may also be important. Diffusion research indicates that innovations are likely to be adopted by individuals who are of high status, and who are highly educated (Rogers & Shoemaker, 1971). Within education, it is frequently thought that younger teachers are more innovative than older teachers. In addition to the variables of *status* (measured by *father's education level*),

Table 12. School Input (Staff-Characteristics) Variables: Operational Definitions[a]

Variable	Operational definition
1. Professional experience/behavior characteristics	
Teaching experience	Mean number of years of teacher experience
Professional reading	Mean number of professional books read in last year
Staff modernity (cosmopolitanness)	Mean modernity of states in which respondents have worked as educators[b]
2. Personal characteristics	
Educational level	Percentage of staff with more than a Bachelor's degree
Father's educational level	Percentage of staff who have fathers with some college education
Age	Mean age of staff
Percentage male	Percentage of staff who are male

[a] Staff-characteristics variables are individual items. Reliability coefficients were not, therefore, calculated.

[b] Modernity of school district experience scores were computed by assigning a state modernity score (see Herriott & Hodgkins, 1973, Table 4-2) to each school district that a staff member had worked in, averaging those scores for each respondent after weighing them by the number of years of experience in each district, and then computing the mean for the school as a whole by averaging staff scores.

educational level, and *age,* we also included a variable measuring the *percentage of male teachers* in the school. (The variables and their operational definitions are presented in Table 12.)

A correlation matrix of all independent variables is presented in Appendix C.

THE EFFECT OF SYSTEM CHARACTERISTICS ON IMPLEMENTATION

In order to determine the relative importance of individual structure, culture, and input variables in explaining implementation of planned change, we conducted separate, stepwise multiple regressions of our three measures of implementation on each of the three groups of independent variables. (For the benefit of those of our readers who are not familiar with regression analysis, we have provided a footnote that briefly explains this statistical technique and offers some guidance in reading the tables of results that follow.)[8]

[8]Multiple regression is a technique used to analyze the relationship between a dependent variable (for instance, quality of implementation) and a set of independent variables (for instance, our 10 school-structure variables). Quality scores in schools ranged from 2.5 to 57; to what extent can this variation ("variance") be explained (or predicted) by these schools' scores on the 10 structure variables?

The results of the regression of quality on the 10 structure variables appear in Table 13 as the second column of numbers (standardized regression coefficients). We have entered the standardized regression coefficients for five variables (e.g., formalization) because these variables were very weak predictors of quality of implementation. Of the five coefficients that do appear, four (those with asterisks) are strong predictors of quality either positively or negatively (e.g., classriom autonomy); in the latter case, the minus sign merely means that *low* classroom autonomy is somehow associated with *high* quality of implementation.

At the foot of the table the reader will see two final entries. Multiple R^2 represents the percentage of variance in the dependent variable (quality) explained by the five independent variables shown—a sizable 50%. (The remainder of the variance in quality is due to other causes.) In view of the large number of variables and the small sample size (45 schools), we also present a more conservative estimate of the variance predicted—adjusted multiple R^2.

Multiple regression is useful for two reasons. Given a set of independent variables, it helps us weed out the weak predictors and concentrate on the strong ones. Furthermore, the figures for multiple R^2 give us an idea of how useful the entire set of variables is for our purposes—understanding the variability of implementation among ES schools.

The Effect of Structure

Table 13 presents the standardized regression coefficients for those structure variables that contributed at least 2% to the explained variance of the total scope-of-implementation score or to either of its dimensions—quality and quantity. The structure variables as a group predict a relatively large percentage of the variance in each of the three measures of implementation. The percentage of explained variance ranges from 50 in the case of quality to 33 in the case of quantity. Even when the adjusted multiple R^2s are examined, we find that with four or five structure variables we are able to explain at least one-fourth of the variance in our measures of implementation.

Two structure variables stand out from the other eight as having consistently significant relationships with both the total implementa-

Table 13. Standardized (Beta) Coefficients for the Regression of Each of Three Measures of Implementation on School Structure Variables (N = 45)

	Implementation measure		
School structure variable	Total scope	Quality	Quantity
Classroom autonomy	–.16 (2)	–.25[a] (4)	—
Complexity	–.22 (5)	—	–.26 (4)
Formalization	—	—	—
Individualized technology	.29[a] (4)	.22 (5)	.37[a] (2)
School level	—	.34[a] (3)	—
School size	.41[a] (3)	.28[a] (2)	.40[a] (3)
School-board authority	—	—	—
Superintendent authority	.39[a] (1)	.48[a] (1)	.31[a] (1)
Principal authority	—	—	—
Teacher authority	—	—	—
Multiple R^2	.43	.50	.33
Adjusted multiple R^2	.36	.43	.27

[a]Nonstandardized regression coefficient is at least twice its standard error.

Note: For each of the three regressions, beta coefficients are presented for only those variables that would increase the stepwise multiple R^2 by at least 2%. The sequence of variable entry into the regression equation was unforced. (Order of entry is presented in parentheses below the associated coefficient.)

tion score and the dimensions of quality and quantity. These are the level of the superintendent's authority in decision making and the size of the school, both of which are positively related to implementation. While the finding that size is related to innovation in organizations is far from novel (cf. Kimberly, 1976), the importance of the role of a strong chief administrator (in this case, the superintendent) is somewhat more surprising. This is particularly the case since neither principal nor teacher authority variables contributed to the explanation of change. Many theorists of planned change argue that organizational innovation is facilitated by decentralized authority structures (Bennis, 1966; Zaltman *et al.*, 1973). Our data, however, indicate that, for comprehensive change of the type envisioned by the designers of the ES program, not only does centralization appear to facilitate implementation of change, but giving teachers significant decision-making power within the classroom (our classroom autonomy measure) may actually inhibit organizationwide innovations.[9]

Another variable which appears to be important in explaining implementation is the use of individualized teaching technologies. Some schools had already had experience in using teaching methods which involved differentiated activities within the classroom; the implementation of new curricular and structural innovations (many of which, within the programs designed by the 10 districts, involved a greater emphasis on individualized methods) may therefore depend, in part, upon the extent to which a school has already adopted such approaches to classroom instruction.

In many studies of organizational change, complexity is found to be positively associated with innovation (Hage & Aiken, 1970; Zaltman *et al.*, 1973). Our data indicate, however, that for rural schools, complexity is not among the most significant predictors of implementation and, furthermore, it tends to be negatively related to change. This finding is not totally anomalous; several other studies have found either insignificant or negative relationships between complexity and change (Hage & Aiken, 1970; Louis, 1977). The inconsistency of results

[9]This finding may be confounded by differences between the organizational/administrative structures of urban and rural school districts—that is, there may be intermediate bureaucratic levels in larger school systems which are not found in the Rural ES districts. (For a more extensive discussion of the impact of centralization and participation in change, see Louis & Rosenblum, 1977.)

across studies indicates that the concept of complexity may require additional theoretical and empirical specification. One possible interpretation of the negative finding in this case is that specialists attached to the various schools may tend to generate their own programmatic priorities which are not always consistent with the objectives of a districtwide change program.

It is interesting to note that within our sample, the variable indicating whether a school is a primary or secondary school is significant for only one measure of implementation—quality. The folklore of education often assumes that primary schools are considerably more innovative than secondary schools, in part because of their simpler organizational structure, and in part because of their greater emphasis on teaching techniques as compared to specialized subject matter. Recent studies have corroborated this assumptions (Berman & Pauly, 1975). Our data, however, suggest that secondary schools are more likely to implement programs which are *significantly different* from previous programs (quality of change) than are elementary schools. Based on our knowledge of the content of the innovations with the 10 districts, we believe this finding may be explained quite simply: elementary schools were more likely to be involved with innovations which modified existing core curricula, such as reading or other basic skills. Secondary schools, on the other hand, were generally involved in implementing totally new programs or courses which were distinct from the previous educational offerings: new career counselors, specialized vocational offerings, new media centers, and so forth. (However, our findings on the continuation of the ES activities, presented in Chapter 9, indicate that, while these noncore innovations were, in fact, more "different," they were also more likely to become dramatically altered or even eliminated after the funding period was over.)

A tentative image of the innovative rural school may be drawn from the findings presented in Table 13. Such a school is a relatively large one, which in the case of the ES rural districts generally means over 200 students. Despite its size, it has a relatively undifferentiated administrative structure and few specialists attached to the school itself. The superintendent, on the one hand, has a great deal of influence in this school and exercises strong authority in program management. The teachers, on the other hand, have relatively little autonomy, which would suggest that the "professionally oriented"

Table 14. Standardized (Beta) Coefficients for the Regression of Each of Three Measures of Implementation on School Culture Variables (N = 45)

School culture variable	Implementation measure		
	Total scope	Quality	Quantity
Change orientation	—	—	—
Collegiality	.83[a]	.43[a]	.80[a]
	(1)	(2)	(1)
Goal differentiation	—	—	—
Goal discrepancy	—	—	—
Level of tension	—	—	.20
			(2)
Frequency of disputes	.24	.25	—
	(5)	(3)	
Morale	−.39[a]	−.38[a]	−.27[a]
	(2)	(1)	(3)
Pupil autonomy orientation	.19	—	.20
	(4)		(5)
Perception of problems	.31	.19	.28
	(3)	(4)	(4)
Multiple R^2	.42	.29	.40
Adjusted multiple R^2	.34	.22	.33

[a]Nonstandardized regression coefficient is at least twice its standard error.

Note: For each of the three regressions, beta coefficients are presented for only those variables that would increase the stepwise multiple R^2 by at least 2%. The sequence of variable entry into the regression equation was unforced. (Order of entry is presented in parentheses below the associated coefficient.)

school with an active staff that participates in curricular decisions may not be the most open to districtwide, comprehensive educational change.

The Effect of Culture

School culture variables as a group appear to have a slightly weaker impact upon implementation than structure variables (Table 14). The adjusted multiple R^2s range from .22 for quality to .34 for the total scope score. Three varaibles fail to explain even 2% of the variance in any of the measures of implementation.

In contrast, collegiality and morale explain significant amounts of variance in all three of our measures of implementation. On the one hand, high implementing schools are characterized by high collegiality. On the other hand, schools marked by high dissatisfaction among teachers with their individual influence on educational decisions (low morale) also seem to implement ES changes at a greater level. To this finding may be linked the positive association of level of tension and frequency of disputes with various measures of implementation.

These results lend support to two general approaches to change that are often viewed as contradictory. Conflict theories of change emphasize the need for unrest in the system at the level of the individual as well as at the level of groups interacting together (Coser, 1956). Functional theories of change tend to stress cohesiveness and cooperation among groups as a prerequisite for smooth, nonrevolutionary change (Parsons, 1951). Our preliminary investigations suggest that both conflict and cooperation need to be present in order to maximize the smooth implementation of new programmatic efforts in schools. Our findings point to the cohesiveness of the staff in the face of organizational and personal dissatisfactions as a factor facilitating change.[10]

One final result of these regressions at first seems paradoxical—our variable "change orientation" turns out to be a poor predictor of all three measures of change. (Even when the less stringent test of contributing 1% to the explained variation was applied, change orientation entered only the regression using quality of change as the dependent variable.) This finding stands as a cautionary note to the researcher or practitioner who may assume that progressive attitudes among staff members will necessraily enhance organizational-change efforts. But our studies of the Rural ES districts and their projects are peppered with instances where teachers who perceived themselves to be innovative resented any attempt to tell them how to change their own teaching procedures (Herriott & Gross, 1979, Chapters 4–8). Where the innovation is compatible with the orientations of the innovative staff it may be easily accepted. Where it is not, however, the innovative staff may be more likely to reject it outright than the noninnovative staff.

[10]Morale is not a well-articulated concept in most organizational research, and is, therefore, frequently measured in radically different ways. Perhaps the most common way of measuring morale is through attitudinal items asking about job satisfaction. Our measure of morale, however, revolved around the notion of satisfaction with the level of influence that the respondents felt they could exercise over their working environment—a measure which may be independent of overall job satisfaction.

The Effect of Input

Staff characteristics have a significant predictive relationship with the total scope score and the dimensions of quality and quantity (Table 15). The ability of our input variables to predict implementation varies, however, by dimension. The adjusted multiple R^2 is .51 in the case of quantity and total scope. In the case of quality, on the other hand, only about one-fourth of the variance can be explained by staff characteristics.

Three variables among the group entered all three regression equations in the same order. These three most powerful predictors are: the educational level of the staff, which shows a strong positive relationship with all three measures of implementation; the percentage of staff that is male, which exhibits a strong negative relationship with implementation; and staff modernity, which is positively related to all dependent variables (although the regression coefficient is insignificant in the case of quality).

Table 15. Standardized (Beta) Coefficients for the Regression of Each of Three Measures of Implementation on School Input Variables (N = 45)

School input variable	Implementation measure: Total scope	Quantity	Quality
Age	−.29[a] (4)	−.33[a] (4)	—
Professional reading	.30[a] (5)	.33[a] (5)	—
Teaching experience	—	—	—
Percentage male	−.23[a] (2)	−.24[a] (2)	−.25[a] (2)
Educational level	.51[a] (1)	.47[a] (1)	.49[a] (1)
Father's educational level	−.23[a] (6)	−.22[a] (6)	—
Staff modernity	.23[a] (3)	.24[a] (3)	.15 (3)
Multiple R^2	.57	.58	.29
Adjusted multiple R^2	.51	.51	.23

[a]Nonstandardized regression coefficient is at least twice its standard error.

Note: For each of the three regressions, beta coefficients are presented for only those variables that would increase the stepwise multiple R^2 by at least 2%. The sequence of variable entry into the regression equation was unforced. (Order of entry is presented in parentheses below the associated coefficient.)

The finding that staff educational level is the most important predictor of implementation is not surprising, for it corroborates evidence from many studies of the diffusion of innovations (Berelson & Steiner, 1964; Rogers & Shoemaker, 1971). In addition, a recent study of educational innovation found that aggregate staff educational attainment had a positive effect on technological change in schools (Corwin, 1973). The size of the standardized regression coefficients is, however, somewhat greater than expected, ranging from .47 to .51. The apparently critical influence of staff educational attainment on implementation may be related in part to the fact that the teachers in many of the schools involved in the ES program had extremely low levels of educational attainment by national standards (less than a B.A.), and the very low innovativeness of these schools may inflate the contribution of staff education in the remaining schools.

Somewhat unexpected was the finding that staff modernity is a major contributor to implementation. Previous studies which have examined the relationship between modernity and school outcomes have done so by assessing the modernity of the state in which the school is located as an environmental influence upon the schools (Corwin, 1973; Herriott & Hodgkins, 1973). We have made the assumption that the modernity of the states in which the staff members have taught is a reasonable proxy measure for the complex concept of *cosmopolitanism*. That is, those staff members who have been employed as educators in more technologically advanced states are likely to bring with them to rural areas a set of assumptions about education that are very different from those held by the indigenous rural residents, and more congruent with the innovative objectives of the Rural ES program. This does, in fact, appear to be the case.[11]

Several comments are in order about the predictive power of the variable "percentage of staff male" on the implementation of change. Our first inclination was to believe that this strong relationship was spurious, possibly due to an intervening correlation of maleness with the variable of school level ($r = .36$). However, not only does introduc-

[11]It should also be noted that staffs that score high on the modernity of their professional experience in school districts also tend to have high scores on two other modernity variables that were not included in this equation—the modernity of the states in which they attended college or university, and the modernity of the states in which they attended high school. Thus, there is a consistent tendency for such schools to be populated by nonindigenous educators who have had considerable life experiences outside of the most rural parts of the country.

ing the variable of school level into an equation with "percentage male" fail to change the relationship, but introducing other variables with which maleness is highly correlated (level of autonomy, tension, superintendent authority) also does not remove its effects, although they are somewhat reduced. Interpretations of this finding should be cautious, for it is difficult to think of a theoretical basis on which to make generalizations about maleness and low implementation. However, within a *rural* school context, it is not unreasonable to assume that the dominant culture would reflect, at least to some extent, the stereotypes of rural sex roles, which focus on the male as the authority figure, with the female playing a more nurturing role. Since many of the innovations in the Rural ES program involved changing the authority relationship between teacher and pupil through individualization of curricula, change may have posed a greater threat to male teachers' behavior patterns than to those of females. Other types of innovations may not have shown such a strong association between sex ratios and implementation.

Although no other variable contributes 2% or more to the explained variance of quality of implementation, several other variables show strong, significant impacts on quantity and total. As expected, age is negatively related to implementation and the level of professional reading is positively related to implementation. More surprising, however, is the fact that father's educational level is negatively related to implementation. Evidence from other studies, though, indicates that upwardly mobile individuals are more likely to be innovative (Rogers & Shoemaker, 1971), and it seems reasonable to assume that schools characterized by upwardly mobile individuals (teachers whose fathers have no college education) might also be more innovative.

Predicting Implementation: A Summary of Most Important Variables

In the section above, we have reviewed the structure, culture, and input variables that best predict implementation. In addition, in order to better understand the ways in which system characteristics at one time are associated with organizational change outcomes at a later time, it seems useful to examine an additional set of regressions which include variables from all groups. First, this additional procedure allows us to see which of the many variables that have emerged as significant in previous sections are particularly powerful when pitted against others with which they have not yet competed. Second, this examination of a composite set of variables will alert us to the relative predictive power of the three groups by indicating the degree to which variables from each group are likely to enter early into a stepwise regression procedure.

In Table 16 we present the results of three stepwise regressions of total scope of change, and quantity and quality on all of the structure, culture, and input variables. The table selectively presents only the five variables that entered first into each regression equation.

Two features of this table are particularly outstanding. First, there is almost no overlap between the group of variables that best predicts the quantity of change and that which best predicts the quality of change. Second, among those five variables that are the best predictors of quantity of change there is a mixture of input, structure, and culture

Table 16. Standardized (Beta) Coefficients for the Regression of Three Measures of Implementation on Structure, Culture, and Input Variables (N = 45)

	Implementation measure		
Variable	Total scope	Quantity	Quality
Education level	.36[a] (1)	.40[a] (1)	
Individualized technology	.38[a] (2)	.26[a] (2)	
Collegiality	.45[a] (3)	.51[a] (3)	.28[a] (5)
Size	.34[a] (4)		.25[a] (2)
Tension	.27[a] (5)	.36[a] (4)	
Professional reading		.25[a] (5)	
Superintendent authority			.44[a] (1)
School level			.37[a] (3)
Classroom autonomy			−.35[a] (4)
Multiple R^2	.59	.63	.52
Adjusted multiple R^2	.54	.58	.46

[a]Nonstandardized regression coefficient is at least twice its standard error.

Note: Only the five variables that entered first into each regression equation are presented. (Order of entry is presented in parentheses below the associative coefficient).

factors. In the case of quality, on the other hand, four of the five most important predictors are structure variables.

A school that has achieved a broad penetration of innovative programming into its daily schedule, and that has achieved comprehensiveness in terms of the variety of staff and pupils affected, is one which is characterized by a strong professional staff which is highly educated and which tends to keep up with current ideas in education through professional reading. The staff members are also mutually supportive of one another: they interact frequently, and feel that they share many values with regard to education. Despite the high levels of mutual support and collegiality, there is a tendency for there to be tension within the school system: teachers often disagree with one another, and with their administrators. The dynamics of this closely knit group with its high levels of disagreement raises the image of a familylike environment which may be associated with the small size of the organizations and communities in which the groups are located. Finally, the school that scores high on quantity of change is also likely to be one that has recently experimented with individualized instruction—in other words, a school that has at least some experience with the implementation of innovations.

The school that has been able to introduce innovative programming that has actually made significant differences in the school's structure and curriculum is one which is characterized by its large size and high levels of collegiality. In addition, the school that scores high on quality of change is one which is highly influenced by policies that emanate from the district central office, and one in which teachers have relatively little autonomy in their own classrooms concerning pedagogical and scheduling decisions. Finally, secondary schools are more likely to introduce innovative programs that are significantly different from previous offerings than are elementary schools.

Summarizing the Results of the Initial Regressions

Our initial regression analyses raise a number of important issues. First, we find that a number of the variables representing structure (school size, superintendent authority, and individualized technology), culture (morale and collegiality), and input (educational level, percentage male, and staff modernity) are highly related to each of the two dimensions and to total scope of implementation. Second, we find that all three sets of variables result in relatively high adjusted multiple R^2s; that is, they explain a relatively large percentage of the variance in each of the dimensions and in the total scope score (from 22 to 57%).

The percentage of the variance explained by each group of variables, when coupled with our knowledge that there are at least modest correlations among many of the independent variables, leads inevitably to an exploratory question: How much of the explanatory power of structure, culture, and input variables reflects unique contributions of each, as opposed to joint contributions in explaining variations in implementation? Further, do interactions among structure, culture, and input variables make significant additional contributions to the explanation of implementation of innovation in schools? These questions are extremely important in exploring a theory of planned change in schools, and are the focus of Chapter 7.

7

Further Exploration of Implementation in Schools

Chapter 6 revealed not only that implementation varied among schools in the 10 ES districts, but also that three variable groups (derived from the general systems framework which had guided our research) predict with considerable accuracy the degree to which implementation occurred. In this chapter, our intention is to explore in greater detail some empirical and theoretical issues surrounding implementation of change in schools. Specifically, the following questions are addressed:

1. How do the different groups of independent variables—structure, culture, and input—relate to each other and interact in ways that predict implementation?
2. How does a particular type of *system linkage,* that is, linkage through the bureaucratic authority structure, affect program implementation?
3. What can we say, more generally, about the impact of system linkages on the implementation of change in schools?

Although at first glance these questions may seem rather diverse, they emerge quite naturally from the theoretical concerns that were outlined in Chapters 1 and 2. An exploration of them will further the ultimate objective of this volume—to enrich existing theoretical approaches to organizational and educational change, and to increase understanding of how human interventions and policies at the federal and local levels may affect the outcome of planned-change programs.

The first question addresses itself to the need to integrate alternative theories of change, which we outlined in Chapter 2. We have already shown that the three independent-variable groups used in our analysis—culture, structure, and input—are all strongly related to implementation. This finding does not, in itself, answer the controversial question of which variable group or change theory is the "best" theoretical explanation for change. In order to address this question head-on, we must pit the different variable groups against each other as competing explanations, and determine which if any is the most important or decisive predictor. In order to explain the results of such analysis, it is also important to examine the interrelationships among the variables in the different groups to determine why or how they contribute independently and/or jointly to organizational change.

The second question addresses a fundamental issue in change theory: the use of control structures to facilitate change. The dominant ideology in most management theory is that participation (which implies a close linkage between administrative and lower level decision making) is necessary in order to achieve widespread or major organizational changes. However, this assumption is far from fully supported at the empirical level, particularly in the educational sphere.

In addressing the question of the impact of system linkage, we look both at the degree of linkage within individual schools and at the degree to which schools are linked to one another at the district level. Through this analysis, we hope to add to our understanding of the degree to which this relatively new elaboration of systems theory is a powerful empirical approach to the explanation of educational innovation.

While this list of questions is clearly selective, we believe that the three general theoretical issues that underlie our choices represent major and unresolved debates in the field of educational and organizational change.

COMPARING THE CONTRIBUTIONS OF STRUCTURE, CULTURE, AND INPUT VARIABLES TO SCHOOL CHANGE

The regressions presented in Chapter 6 demonstrated the strong predictive power of the variable groups of structure, culture, and input. We felt that it was important to explore further the degree to which each of these variable groups represented a valid alternative explanation for implementation. To do this, we again performed regressions of each of the three measures of implementation; however, the regressions were performed with two variable groups at a time (structure and culture, structure and input, and culture and input). We then performed a "commonality analysis" by computing the *unique* contribution of each paired variable and the *joint* contribution of both to the percentage of variance explained.[1] Table 17 displays the results of the nine regressions performed.

Turning to the "bottom lines" first, the table indicates that between 40% and 72% of the variance in the implementation scores can be explained on the basis of intraorganizational variables alone. The size of these multiple R^2s is much higher than is generally found in studies that attempt to predict implementation of change in organizations on the basis of similar variables. Baldridge and Burnham (1975), for example, are able to predict only about 30% using both intraorganizational variables and an additional cluster of variables reflecting the environment of the school. Using both district and school level variables, Deal *et al.* (1975) explained only 23% of the variance in school adoptions of team teaching and differentiated reading instruction. The Rand Change Agent Study (Berman & Pauly, 1975) explained only between 17% and 35% of the variance in their several dependent measures of program implementation. Because our multiple R^2s are so much higher than these, it is worth commenting on some possible reasons for the difference.[2]

[1]We analyzed the commonality of the structure, culture, and input variable groups as they relate to the three dependent variables as follows:

$$X_u = T - Y$$
$$Y_u = T - X$$
$$XY = T - (X_u + Y_u)$$

where X_u and Y_u are unique contributions of variable groups in their respective regressions, XY is the common contribution, and T is the total variance explained by both sets. For these regressions, we used only structure, culture, and input variables that met the criterion of contributing at least 2% to the multiple R^2 of the regressions presented in Chapter 6.

[2]Hage and Aiken (1967) were able to explain 55% of the variance in "number of program changes." However, their results should be interpreted with some caution since they had only 16 cases, and entered seven variables into the regression analysis. This points up the problem we faced of "shrinkage" in the R^2 when degrees of freedom are used up by entering a large number of variables. We have discussed the unadjusted R^2, following existing conventions in the sociological literature. However, the adjusted R^2s in Table 17 indicate that "shrinkage" due to diminishing degrees of freedom is not a major problem, as the adjusted R^2s are still higher than those generally found in the literature.

Table 17. Proportion of Variance in Each of Three Measures of School-Level Implementation Explained Uniquely and Jointly by Various Two-Variable Groups (N = 45)

Variable groups	Implementation measure		
	Total scope	Quality	Quantity
Structure and culture			
Unique contribution of structure variables	.19 (.15)	.18 (.14)	.16 (.13)
Unique contribution of culture variables	.17 (.13)	.07 (.01)	.24 (.19)
Joint contribution of culture and structure	.25 (.21)	.22 (.21)	.17 (.14)
Total multiple R^2	.61 (.49)	.47 (.36)	.57 (.46)
Structure and input			
Unique contribution of structure variables	.08 (.03)	.14 (.12)	.07 (.01)
Unique contribution of input variables	.15 (.10)	.05 (.01)	.22 (.16)
Joint contribution of structure and input	.35 (.33)	.26 (.21)	.26 (.26)
Total multiple R^2	.58 (.46)	.45 (.34)	.55 (.43)
Culture and input			
Unique contribution of culture variables	.19 (.19)	.12 (.06)	.24 (.23)
Unique contribution of input variables	.30 (.28)	.11 (.07)	.31 (.31)
Joint contribution of culture and input variables	.22 (.15)	.17 (.16)	.17 (.10)
Total multiple R^2	.71 (.62)	.40 (.29)	.72 (.64)

Note: The adjusted multiple R^2 appears in parentheses. All coefficients are significant at or below the .05 level.

First, the high levels of association found represent an initial validation of the scope-of-implementation construct. The research was initiated with the premise that developing a better understanding of change required knowing in some detail what the change was. By differentiating the concept and operationalizing several distinct measures of change, we feel that we have reduced measurement error and increased our ability to find important relationships.

Second, we must note that we are looking at the implementation of particular, locally planned changes that were designed and put into practice within a district. Many other researchers have been concerned either with the adoption of one or two innovative programs that are being diffused throughout the United States or with more general "program change." (See Downs & Mohr, 1976, for a critique of this approach to measuring innovation.) The greater specificity of our dependent measure results, in large part, from the nature of the Rural ES program.

Table 17 also provides support for the assumption that we need to build conceptual bridges between approaches that emphasize cultural, structural, and input explanations of organizational change. No one variable group is preeminent; rather, we find that each of the variable sets that are alternative explanations of change makes a unique contribution to the variance in implementation of ES innovations. The relative contribution of each variable group to the explanation of implementation varies depending on the other group with which it is paired. In the case of structure and culture, the two groups contribute uniquely to our understanding of total scope to a roughly equivalent degree. When input is paired with structure or culture, it tends to be somewhat more powerful. Looking at the dimensions of quality and quantity, however, we see a slightly different pattern. Quality of change is, apparently, best explained by structure variables; when they are paired either with culture or with input variables, structure variables show somewhat greater unique predictive power. Quantity, however, is best explained by input. These findings are extremely important, for they suggest why all three types of variables may be of theoretical and practical significance.

In the absence of a staff which has certain key characteristics—upward mobility from family of origin, a relatively cosmopolitan professional background, and a high level of educational attainment—the pervasiveness and extent of the involvement of staff and students in innovative activities will be low. The characteristics of the staff do not appear to affect the quality of innovation—the degree to which it represents a major change from previous practices—but do have a significant impact upon the degree to which innovative practices represent substantial portions of the educational activities of the school. This finding suggests that it is important to reassess the way in which we talk about teachers' "resisting" innovation.

Resistance to innovation by certain types of teachers is not a simple phenomenon. Our data suggest that relatively significant changes can be introduced into schools with different types of staffs (as long as they are relatively well educated as a group). Nevertheless, staff characterisitcs may affect the degree to which the staff will be willing to devote significant amounts of their time or high percentages of pupil time to the innovative activity. For the planner of change, this fact suggests that staff characteristics should be taken into consideration in determining whether the innovation should be introduced gradually, through demonstration classrooms or other similar techniques, or whether it may be introduced more rapidly and on a less experimental basis.

Unless there is a supportive school culture, planned innovations may be isolated in a limited number of classrooms or may involve a very small percentage of the normal school day. The innovations will not necessarily disappear, but they will be confined to those individual teachers who are willing to use them or will be minimized in terms of their impact upon the total teaching environment. Given the centralized nature of the planned-change activity being studied (district-wide planning and implementation of new programs), isolation of the innovation is one technique that may be used to acknowledge a demand from high levels of authority to implement change while, in fact, allowing it to have only minimal impact upon organizational operation.

In the case of the quality of innovation, or the degree to which it represents a genuine change of activities, structural features such as centralized district leadership, low levels of classroom autonomy, and the presence of individualized teaching technologies may be required as a prerequisite to the designing of programs that depart significantly from existing practices. For example, in a school where teacher control over the classroom is high, it may be difficult to introduce an innovation that will actually be implemented in a similar fashion across classrooms, thus producing an innovative school-wide program. Such a structure may produce so much individual adaptation at the classroom level that observing the innovation as a similar activity across classrooms may be impossible. Furthermore, small schools may simply not have the staff resources to implement a program that is very different from existing practices, even when they are provided with new materials and occasional training support, as was generally the case for the ES projects.

Equally important to the finding that culture, structure, and input tend to contribute differentially to the explanation of the quality and quantity of change is a final finding that emerges from Table 17. In all nine regression equations, the joint contribution of the paired variable sets is high, ranging from .17 to .35. In terms of percentage, the joint contribution of variable groups explains from 24% (joint contribution of culture and input to quantity) to 60% (joint contribution of structure and input to total scope) of the total explained variance. This finding suggests, as noted above, that in order to advance our understanding of change, we must explore the ways in which the different variable groups interact in their relationships to the implementation of change.

It is interesting to note that the percentage of the variance that is not uniquely attributable to a single variable group is, on average, highest in the case of the structure/input pair and lowest in the case of culture/input. This is contrary to what might have been anticipated, since it is frequently assumed that the culture or climate of the organization is more dependent on the characteristics of the people who inhabit it than upon the structures that characterize it. This finding suggests that we should expect to find greater explanatory power for interaction terms in the case of structure/input and less in the case of culture/input—a topic to which we now turn.

EXPLAINING JOINT CONTRIBUTIONS OF STRUCTURE, CULTURE, AND INPUT

How can we explain the high levels of joint contribution of the three independent-variable groups to the variance in implementation? Two alterantive explanations seem plausible. The first concerns the measurement and conceptual clarity of the three variable groups. Our classification of variables into structure, culture, and input categories was originally made by a test of face validity based on theoretical support in the research literature. However, particularly with regard to the structure and culture variables, which were measured largely through attitudinal batteries on questionnaires, it seemed useful to determine whether, in fact, the variables reflect separate and unique constructs of "structure" and "culture." In order to test for the conceptual distinctiveness of our variables, we submitted them to a cluster analysis. This procedure resulted in six clusters which very closely paralleled the structure and culture categories. The six clusters, however, failed to have very much explanatory power in predicting variance in implementation, and therefore were not used further in the analysis.[3]

The second possible explanation was that our high levels of common contribution occurred largely as a result of *interactions* among variables in the three independent groups. Pronounced interaction effects would explain the instances where there was a high level of common contribution of two groups in an implementation measure. The major approach which was used in the analysis was to explore the interaction effects of variable pairs.

Interaction Effects between Variable Pairs

In the behavioral sciences, conditional relationships between independent variables have frequently been found to be critical in predicting a dependent variable. For example, it is now taken almost for granted that sex and race interact with one another in their impact upon achievement motivation (Horner & Walsh, 1974) and that there are aptitude–treatment interactions in the educational process (Bracht, 1970). Studies of organizational behavior, on the other hand, have only rarely attempted to look for systematic interaction effects (e.g., Herriott & Hodgkins, 1973), although the development of contingency approaches to management theory has stimulated development in this analytic area (see, e.g., Bons & Fiedler, 1976; Fiedler, 1972).

Although the findings described above suggest that it may be fruitful to examine interaction effects *within* structure, culture, and input variable groups as well as among them, this analysis limits its examination to interactions *among* groups of variables. In order to locate significant interactions, a number of steps were taken.

First, it was decided to limit the examination of interactions to those variables that appeared to have consistently strong predictive relationships with the total scope of implementation, and with the dimensions of quality and quantity. Four structure variables (superintendent authority, classroom autonomy, school size, and individualized technology), four culture variables (collegiality, morale,

[3]For this reason, we do not include a description of the cluster analysis and its results in this chapter. However, the interested reader can refer to Jastrzab, Louis, and Rosenblum (1977) for a description of cluster analysis, the results of the cluster procedure, and the regression analyses performed.

perception of problems, and level of tension), and three input variables (educational level, percentage male, and staff modernity) were selected.[4]

Interaction terms were then computed by multiplying the score for each variable pair.[5] This procedure yielded 40 interaction terms. Because of the large number of main and interaction terms and the relatively small number of schools in our sample, it was decided to examine interactions in the following ways. For each of three variable pairs—structure/culture, structure/input, and culture/input—separate regressions were conducted for the dependent variable of total scope and its dimensions of quality and quantity. The main terms were entered in a block on the first step, and the interactions between the main terms were allowed to enter an unforced stepwise procedure on the remaining steps. The results of these regressions are presented in Tables 18, 19, and 20.

It was anticipated that, because of the limited number of degrees of freedom available after the entry of the main terms, and the relatively large multiple R^2 associated with the main terms, the interaction variables would have only a slight effect upon the adjusted multiple R^2. However, the regression results indicate that the addition of interaction terms to the equations has a strong impact upon the explanation of implementation.[6]

Where structure and culture variables are paired, the addition of

[4]The selection was made on the following basis: using the three dependent variables, nine regressions were computed. Three involved entering all structure and culture variables stepwise, three only structure variables, and three only culture variables. Structure and culture variables that contributed at least 2% to the explained variance in at least three of the six regressions where they were included were classified as having consistently strong predictive power. Several colleagues have noted that they were concerned about the fact that educational level was not included among our group of powerful variables, and wondered whether further exploratory analysis might have revealed it as a significant factor in explaining change patterns. We responded to this concern with additional attempts to look for impacts of whether a school was an elementary or secondary school (Jastrzab *et al.*, 1977), but were unable to locate any results which systematically illuminated our more powerful findings.

[5]In order to avoid a problem of multicollinearity between main and interaction terms, the mean was subtracted from each score prior to its multiplication.

[6]It is important to emphasize that the adjusted multiple R^2 continued to increase through each step of the regression rather than declining, as it would if the "shrinkage" due to lost degrees of freedom outweighted the increased fit obtained through the addition of new variables. While the number of variables entered is large compared to the degrees of freedom, the increase allows us to be confident that we are not, in fact, "over-predicting" the equation.

interaction terms to the main terms in the regression equation increases the adjusted multiple R^2 by at least 11% (Table 18). In the case of structure/input interactions, increases range from a low of five percentage points in the regression using total scope as the dependent variable, to a high of 15 percentage points where quality is the dependent variable (Table 19). Increases are smaller in the case where culture and input are paired, ranging from no increase for total scope (no interactions meeting our criteria for entrance into the equation) to a high of 8% for the equation with quantity (Table 20). The lower impact of interactions on the culture/input equation is, however, expected

Table 18. Standardized (Beta) Coefficients for Main and Interaction Terms in the Regression of Each of Three Measures of Implementation on Selected Structure and Culture Variables (N = 45)

Structure or culture variable	Implementation measure		
	Total scope	Quality	Quantity
Main terms			
Superintendent authority	.19	.33[a]	.23
Classroom Autonomy	.09	−.03	−.01
Size	.48[a]	.27	.37[a]
Individualization	.37[a]	.23	.54[a]
Collegiality	.60[a]	.24	.54[a]
Morale	.16	.17	−.19
Problems	.41[a]	.25	.40[a]
Tension	−.01	−.10	.24
Interaction terms			
Size/tension	.67[a]	.35[a]	—
Size/collegiality	.56[a]	—	—
Individualization/morale	.28[a]	—	—
Superintendent authority/collegiality	.20	—	—
Autonomy/collegiality	—	−.33[a]	−.32[a]
Autonomy/morale	—	—	−.48[a]
Individualization/tension	—	—	−.30[a]
Multiple R^2			
Main terms	.59	.48	.57
Main plus interaction terms	.77	.59	.69
Adjusted multiple R^2			
Main terms	.50	.36	.47
Main plus interaction terms	.69	.47	.58

[a]Nonstandardized regression coefficient is at least twice its standard error.

Note: For each of the three regressions, all of the main terms were entered prior to any of the interaction terms. Beta coefficients for the interaction terms are presented for only those variables that would increase the stepwise multiple R^2 by at least 2%.

Table 19. Standardized (Beta) Coefficients for Main and Interaction Terms in the Regression of Each of Three Measures of Implementation on Selected Structure and Input Variables (N = 45)

Structure or input variable	Implementation measure		
	Total scope	Quality	Quantity
Main terms			
Superintendent authority	.30[a]	.66	.16
Classroom autonomy	−.09	.11	−.23
Size	.11	.20	−.01
Individualization	.16	−.15	.23
Percentage male	−.21	−.18	−.19
Education level	.44[a]	.08	.54[a]
Staff modernity	.21	.15	.14
Interaction Terms			
Size/percentage male	.22[a]	—	—
Individualization/Education	—	.45[a]	—
Individualization/percentage male	—	−.38[a]	—
Autonomy/percentage male			−.28[a]
Size/Education	—	—	.25[a]
Multiple R^2			
Main terms	.55	.45	.49
Main plus interaction terms	.50	.64	.59
Adjusted multiple R^2			
Main terms	.46	.34	.41
Main plus interaction terms	.51	.49	.50

[a] Nonstandardized regression coefficient is at least twice its standard error.

Note: For each of the three regressions, all of the main terms were entered prior to any of the interaction terms. Beta coefficients for the interaction terms are presented for only those variables that would increase the stepwise multiple R^2 by at least 2%.

since the level of unique contributions of each of the two variable sets was much higher in this case than for the other pairs (see Table 17).

One interesting feature of these analyses is that different interaction terms tend to emerge as critical for total scope, quantity, and quality. This finding tends to support the belief that quantity and quality are, in fact, quite different measures of the outcome of innovative programs and school activities.

In summary, 15 interaction terms emerge as significant in at least one equation. These significant interactions tend, however, to form a discernible pattern, involving a somewhat more limited set of variables: (1) Four significant interactions include size as one of the terms; (2) five significant interactions include percentage male as one of the

Table 20. Standardized (Beta) Coefficients for Main and Interaction Terms in the Regression of Each of Three Measures of Implementation of Selected Culture and Input Variables (N = 45)

Culture or input variable	Implementation measure		
	Total scope	Quality	Quantity
Main terms			
Collegiality	.56[a]	.44[a]	.64[a]
Morale	.15	.30[a]	.13
Problems	.21	.20	.13
Tension	.09	−.07	.23
Percentage male	−.28[a]	−.19	−.24[a]
Education level	.44[a]	.37[a]	.51[a]
Staff modernity	.29[a]	.06	.30[a]
Interaction terms			
Problems/Modernity	—	.32[a]	—
Morale/percentage male	—	—	.38[a]
Collegiality/percentage male	—	—	.32[a]
Multiple R^2			
Main terms	.59	.37	.64
Main plus interaction terms	—	.44	.72
Adjusted multiple R^2			
Main terms	.52	.25	.57
Main plus interaction terms	—	.32	.65

[a]Nonstandardized regression coefficient is at least twice its standard error.

Note: For each of the three regressions, all of the main terms were entered prior to any of the interaction terms. Beta coefficients for the interaction terms are presented for only those variables that would increase the stepwise multiple R^2 by at least 2%.

terms; (3) four significant interactions include individualized technology as one of the terms; and (4) four significant interactions include an authority-related variable (superintendent authority or classroom autonomy) as a term. Only one significant interaction, that between staff modernity and perception of problems, cannot be classified into one of these four groups.

Discussion of Interaction Terms

Although the regressions are useful in determining whether interaction terms can significantly improve our ability to predict implementation of innovations in schools, the regression coefficients themselves do not reveal in what way the variables are interacting in their relationship with the dependent variables. In order to examine the interaction terms further and to interpret them, each of the variables included in an interaction term was dichotomized at the median, and the means of the dependent variables in each of the cells resulting from the pairing of dichotomized variables were examined. The results of this decomposition are presented in Tables 21 (for total), 22 (for quality), and 23 (for quantity). We present our interpretations of terms that showed interaction effects for one or more of the measures of implementation.

Interactions between Size and Other Variables

Size, which is one of the most powerful predictive variables of those included in this analysis, is a component of four significant interaction terms: size/tension; size/collegiality, size/percentage male, and size/staff educational level. It is important to emphasize that these interactions do not "explain away" the effects of size. In several cases, size retains a significant, independent standardized regression coefficient even when the interaction term is entered. Nevertheless, the interactions do moderate the influence of size, indicating that explanations of change must not be predicted simply on, for example, the equation of largeness with slack resources.

The *size/tension* interaction (Table 21) suggests that tension in a school may either facilitate or impede planned change, depending upon the size of the school. Small schools characterized by high levels of tension between role partners are significantly less likely than other schools to be high implementers, while large schools with high levels of tension are significantly more likely to be high implementers. This finding suggests that the question of whether tension is "healthy" may be resolved, in part, by examining the context in which tension occurs. Tension within small work groups, which are often characterized by highly affective relationships among staff members, appears to have negative consequences for the organization's ability to adapt. In large organizations, characterized by more bureaucracy and probably less affect, tension may well serve as a stimulus and an incentive to change.

The interaction between *size and collegiality* (Table 21) also indicates the importance of context. It appears that collegiality has a strong positive impact upon implementation only when it is found in larger schools. Small schools with a high level of collegiality implement no

Table 21. Mean Total Scope Scores for Selected Dichotomized Interaction Terms (N = 45)

		Size: High	Size: Low
Tension	High	54.3 (12)	32.4 (11)
	Low	45.4 (11)	42.7 (11)

		Size: High	Size: Low
Collegiality	High	58.4 (10)	41.4 (13)
	Low	43.6 (13)	32.1 (9)

		Individualization: High	Individualization: Low
Morale	High	47.7 (9)	37.0 (14)
	Low	47.8 (14)	45.0 (8)

		Superintendent authority: High	Superintendent authority: Low
Collegiality	High	53.9 (16)	37.0 (7)
	Low	38.5 (7)	39.0 (15)

		Size: High	Size: Low
Percentage male	High	48.9 (12)	33.3 (8)
	Low	51.1 (11)	45.0 (8)

Note: The *N* for each cell is in parentheses.

more frequently than do large schools with low collegiality. It should also be pointed out that the absence of collegiality in the small school apparently inhibits implementation of innovative programs. This finding is particularly intersting because there is a negative simple correlation (−.21) between collegiality and size. While it may be rare to find a large school with a cohesive staff, the large school that has managed to develop a supportive and close peer environment among its staff members is perhaps more able to take advantage of the "slack resources" associated with its size.

The interaction between *school size and percentage of male staff* (Table 21) indicates a pattern similar to the two interactions discussed above, namely that the "maleness" of the staff is strongly related to implementation only under certain conditions of size. In this case, the percentage of male staff exhibits a powerful negative impact on implementation only in small schools. It is important to point out that this finding does not imply that these small schools are elementary schools, for within our sample, size and school level are not highly correlated ($r = .11$). Rather, we would turn again to the interpretation that maleness and the upholding of traditional, authoritarian culture values are likely to be associated, and that it is in the smaller, more isolated schools within some of the large rural districts involved in ES that these forces are most overtly exhibited.

It should also be pointed out that male staff members in rural schools tend to have serious role-overload problems. In many cases rural male teachers are not only teachers, but administrators, bus drivers, and coaches. Additionally, in most cases in the Rural ES districts, the pay scales for remote schools were insufficient to support a family, and male staff members often held additional jobs in the afternoons and evenings. This type of role overload is inconsistent with the additional time demands for new classroom preparations, meetings, and workshops imposed by the ES planning and implementation efforts.

The final variable that interacts with size is *educational level* (Table 23). Here we find that the correlation between educational level and implementation is more extreme in large schools than in small schools. More specifically, in large schools with fewer staff members with professional degrees beyond the B.A. level, implementation is very low, while in schools of similar size with many better-educated staff members, implementation is particularly high. Apparently there is a "critical mass" effect in which an increase in staff size apparently augments the tendency of less educated staff members to either resist or be unable to effectively deal with the implementation of change, but which creates a supportive group for implementing change in schools chracterized by more highly educated staff members.

It appears that the data provide some support for policy makers who contend that small schools are not the preferred educational structures for rural America. In each case it appears not only that larger schools are more innovative, but that size moderates the positive predictive contributions of other variables such that the greatest implementation occurs primarily in larger schools. In the case of "maleness" we see that the negative associations between maleness and implementation also occur only in the smallest schools. While there may be other reasons for preferring the small, independent school to larger schools in consolidated rural school districts, it seems from these data that small schools are less able to adapt to new programmatic efforts.

Interactions with Individualized Technology

Two interpretable interactions emerged from the regression analysis.[7] The first of these, *individualization/morale* (Table 21), shows a strong negative effect on implementation when the level of individualization early in the project is low and staff morale is high. Since staff morale is strongly correlated with classroom autonomy and teacher authority in decision making, it is clear that this interaction suggests a situation where teachers are happy with their jobs and satisfied with existing levels of individualization. Since greater individualization was a core objective of most ES projects, this interaction may be interpreted as indicative of the way in which low motivation to adopt the innovations affects the level of implementation.

A second interaction occurs between *individualization and staff educational level* (Table 22). Here we find that, under conditions of high individualization, the effects of staff education are augmented. In

[7]While four interaction terms which included individualization as a factor emerged in the regression analysis, two of these did not produce a clear interaction pattern when they were examined in the dichotomized cross-tabular presentations. These two, the interaction of individualization and precentage male (in an equation with quality) and of individualization and tension (in an equation with quality), are therefore not discussed.

Table 22. Mean Quality-of-Change Scores for Selected Dichotomized Interaction Terms (N = 45)

		Autonomy High	Autonomy Low
Collegiality	High	28.7 (12)	40.8 (11)
	Low	35.3 (12)	28.0 (10)

		Individualization High	Individualization Low
Education	High	41.8 (10)	35.6 (12)
	Low	23.6 (13)	28.1 (10)

		Problems High	Problems Low
Modernity	High	42.2 (11)	38.6 (9)
	Low	27.1 (13)	30.7 (12)

		Individualization High	Individualization Low
Morale	High	32.0 (9)	29.7 (17)
	Low	35.7 (14)	39.3 (5)

Note: The *N* for each cell is in parentheses.

other words, the schools that are already rich in resources of staff experience and previous organizational experience with individualized programs have a head start in implementing further changes. The fact that schools with experience in individualized programs but less well-educated staffs are less likely to implement ES efforts at individualization indicates that a lower level of staff education may be a limiting factor in the implementation of greater individualization.

Interactions with "Percentage Male"

The interactions between school size and staff maleness have already been discussed. Another interpretable interaction with the sex ratio among school staff members appeared in the analysis—that between *maleness and collegiality* (Table 23). This interaction shows that schools with a relatively large ratio of male staff and low levels of collegiality are the least likely to implement. Where collegiality is high, however, the effect of maleness disappears entirely. This interaction reinforces the interpretation that it is not some inherent sexual bias against innovation which produces these results, but rather a contextual effect. The overextended rural male teacher may simply have less time to develop the strong collegial relationships that would facilitate the implementation of ES innovations.

Interactions with Variables Related to Authority and Influence in Schools

Two variables related to patterns of authority and influence within schools emerged in several interactions. These include *teacher autonomy* in the classroom and *superintendent authority*.

The interaction between *superintendent authority and collegiality* (Table 21) reveals that the impact of centralized decision making is almost completely negated by the level of peer supportiveness within the school. Where either central district influence or collegiality is absent, the level of implementation is similar, and markedly lower than where both are high. Where the school environment is interpersonally supportive and the administration can mandate change, there are few apparent impediments to instituting change. The image created by this interaction is of the "happy family" school headed by a relatively strong figure.

Examination of the joint effects of *autonomy and morale* (Table 23) reveals an interesting pattern in which the interaction changes the direction of influence of the main terms. Both autonomy and morale are negatively related to implementation. However, where autonomy is low and morale is high—that is, where teachers have little influence but are generally satisfied—implementation is most likely. This combination of morale and actual authority implies a school with a passive but "satisfied" staff which is willing to contribute some extra effort to alter existing methods of instruction. On the other hand, where autonomy is low and the staff is dissatisfied with their control over instructional matters, implementation is low. In such a situation,

Table 23. Mean Quantity-of-Change Scores for Selected Dichotomized Interaction Terms (N = 45)

		Autonomy High	Autonomy Low
Morale	High	46.1 (9)	66.5 (13)
	Low	56.1 (15)	41.7 (8)

		Autonomy High	Autonomy Low
Percentage male	High	49.8 (16)	50.8 (11)
	Low	57.3 (8)	62.7 (11)

		Size High	Size Low
Education	High	65.6 (18)	58.5 (7)
	Low	36.5 (5)	45.3 (15)

		Individualization High	Individualization Low
Tension	High	64.5 (9)	49.3 (14)
	Low	58.6 (14)	45.2 (8)

		Morale High	Morale Low
Percentage male	High	48.0 (46)	53.9 (10)
	Low	58.2 (7)	61.7 (12)

		Collegiality High	Collegiality Low
Percentage male	High	62.2 (9)	43.9 (17)
	Low	63.4 (14)	52.0 (5)

Note: The *N* for each cell is in parentheses.

there seems to be insufficient motivation to participate in the change effort.

A third interaction falling into this group is that between *autonomy and collegiality* (Table 22). Overall, autonomy is negatively related to implementation and collegiality is positively related. However, where autonomy is high, the relationship between collegiality and implementation appears to be weak, and is stronger only in the presence of low autonomy. Like the interaction between superintendent authority and staff collegiality, this interaction reveals that a relatively nonmilitant staff, with low levels of independent decision-making authority in instructional matters but with high levels of peer support, provides a particuarly effective environment for implementation of change.

A more general conclusion may be drawn. The higher the levels of teacher professionalism and teacher autonomy, and the lower the levels of staff collegiality/morale in the school, the less likely it is that major innovations will occur at the school level. Although teacher militants often argue for teacher control over school-level innovations, our data suggest that teacher control may be associated with low levels of school-wide innovation, at least in cases where the stimulus for the innovation comes from outside the school. The data may be revealing a conflict between autonomous teachers and authoritative district administrators over the right of districts to impose new programs within schools, and a resulting antagonism to implementing on the part of teachers.

These findings about the authority structure and levels of implementation raise some additional questions. In particular, our findings challenge the notion that power equalization and decentralization are the most effective means to foster efficient, adaptive organizations. In addition, principal authority does not appear anywhere in our "important predictors," a finding which is at odds with those of other studies. This topic seems to be of such critical importance both theoretically and practically that it calls for further examination of the relationships between patterning of authority and ES outcomes.

HOW DOES THE STRUCTURING OF AUTHORITY AFFECT IMPLEMENTATION?

As we noted in Chapters 1 and 2, a basic tenet of human relations theory is the need for participation by lower level staff members in an organization in order to facilitate change. Furthermore, changes which are highly dependent on members of the organization for implementation and can be easily altered by their actions may require greater centralization than those changes which are straightforward and difficult to subvert. Some educators maintain that critical decisions about planned organizational change must be made by the administration (Bishop, 1961; Brickell, 1964; Heathers, 1967). Even though the question of participation and influence is a complex one, the specific issues to be explored in this section are limited to the following, which are directly relevant to the data which we have:

1. What are the relationships among the influence of various school personnel (superintendents, principals, and teachers) in the process of planning the ES project and the implementation of the planned changes?[8]
2. What are the relationships among the influence of various school personnel on ongoing organizational decisions and the implementation of planned change?
3. How do interactions between the levels of influence of various school personnel over planning and ongoing decision making affect the implementation of planned change?

Analysis and Results

A somewhat unusual pattern emerged from our data with regard to the differential influence of principals both in planning for change and in ongoing decisions (see Table 24). The standardized regression (beta) coefficients for *planning influence* show that the superintendent's involvement facilitates implementation of change, whereas the teacher's role is negligible. The beta coefficient for principal planning influence, however, is moderately high, but with a negative sign. In other words, when the effects of the influence of other actors over planning are removed, an increase in principal influence has negative implications for the successful implementation of the change program. On the other hand, when *general* decision-making influence is

[8]For the purposes of this analysis we introduce three new variables that have not been included up until this point. These are superintendent influence over planning, principal influence over planning, and teacher influence over planning. These variables are derived from a survey question (fall, 1973) which asked respondents to indicate on a five-point scale the level of influence that these and other groups had over the planning of the ES project in their district (see Appendix B).

Table 24. Statistics for the Regression of Total Scope of Implementation on Influence on Planning and on Influence on Decision Making, by Various Actor Groups (N = 45 schools)

Variable	Multiple R	R^2	Adjusted R^2	R^2 change	β
Influence on planning					
Superintendent	.35	.12	.10	.12	.52[b]
Principals	.41	.17	.13	.05	−.28[a]
Teachers	.42	.17	.11	.002	.07
Influence on Decision making					
Superintendent	.51	.25	.25	.25	.52[b]
Principals	.52	.27	.21	.0002	−.07
Teachers	.52	.27	.23	.005	.02

[a]*F* statistic significant at the .10 level.
[b]*F* statistic significant at the .05 level.

examined, neither the level of principal nor teacher influence is associated with implementation. Superintendent influence remains highly correlated.

These results do not, however, take into account a crucial aspect of organizational functioning. In the school change process, actors at the various hierarchical levels do not act independently, but either interact or react to behaviors exhibited by the other actors in the system. Thus, it would be premature to conclude on the basis of the simple correlations and beta coefficients that the key to successful implementation of district-wide planned change is to be found solely in the presence of a strong chief administrator. Rather, we must first examine the possibility that it is the structuring of patterns of influence within the district that determines successful implementation.

In order to make a preliminary examination of the impact of interaction between levels of influence held by superintendents, principals, and teachers, cross-product interaction terms for each of these three role partners were computed for both planning and general decision-making influence. These cross-products were then entered separately into regressions of total scope that include the separate terms for each actor group. For influence over the planning of change, two of the interaction terms contribute significantly to our understanding of implementation (Table 25). The greatest increase is contributed by the interaction between teacher and superintendent influence over planning, with the superintendent/principal interaction also achieving

Table 25. Statistics for Regression of Total Scope of Implementation on Main and Interaction Terms for Influence on Planning and for Influence on Decision Making (N = 45 schools)

Variable	R^2 main terms	R^2 with interaction terms	Increase in R^2	β
Influence on planning				
Superintendent/principal	.17	.23	.05	−.27[a]
Superintendent/teacher	.17	.48	.31	−.67[b]
Principal/teacher	.17	.18	.01	.11
All three interaction terms	.17	.48	.32	—
Influence on decision making				
Superintendent/principal	.27	.27	.00	−.05
Superintendent/teacher	.27	.29	.03	−.17
Principal/teacher	.27	.31	.04	.23[a]
All three interaction terms	.27	.34	.04	—

[a] *F* statistic significant at the .10 level.
[b] *F* statistic significant at the .05 level.

.10 level of significance. For general decision-making influence the interaction between principals and teachers contributed significantly.

The nature of the interactions between influence at various levels in the school hierarchy was further examined by dichotomizing (at the median) each actor group with respect to each type of influence. Mean total scope scores were then examined for those interaction pairs that emerged as significant. Looking first at the data on planning influence, we find that when the superintendent's influence on planning is high, both principal and teacher planning influence are negatively associated with implementation (Table 26). But, under conditions of low superintendent influence, both principal and teacher influence are postively associated with change.

When the effects of high and low influence in the case of principals and teachers are examined, a pattern emerges that is quite different from that found in the case of superintendents. In both cases, where the influence of the school-based actor over planning of change is low, the influence of the superintendent is strongest.

The opposite results appear when we look at the relationships between teacher and principal influence on decision making. Here it appears that when the influence of both is high, implementation is highest.

In other words, influence of principals and teachers appears com-

Table 26. Mean Total Scope Scores for Selected Dichotomized Interaction Terms (N = 45)

		Superintendent authority (planning)	
		High	Low
Teacher authority (planning)	High	81.1 (13)	85.7 (10)
	Low	114.6 (8)	80.3 (14)

		Superintendent authority (planning)	
		High	Low
Principal authority (planning)	High	83.5 (11)	90.2 (9)
	Low	105.7 (10	77.9 (15

		Principal authority (decision making)	
		High	Low
Teacher authority (decision making)	High	92.7 (13)	85.0 (39)
	Low	87.5 (9)	85.3 (14)

Note: The *N* for each cell is in parentheses.

plementary, but between school-based actors and superintendents, no such complementary effect exists. We see instead a pattern that suggests that high influence over planning both within the school and within the central office is the condition which was least likely to produce implementation.

Taken as a whole, the findings suggest a number of conclusions that contradict prevailing assumptions about the importance of participation and influence at various hierarchical levels on the implementation of planned change. They suggest quite strongly that the successful implementation of a *district-wide* change program is most effectively facilitated by a chief administrator who dominates both the planning process and the administrative decision making in the school system. Despite the fact that actual implementation was carried out on the school and classroom levels, high levels of participation of teachers and principals in planning change and high influence of teachers and principals in the general decision-making process in the school and district were not generally associated with higher levels of actual implementation.

The importance of a strong chief administrator is not surprising in light of both empirical findings and change theories that stress the need for strong organizational support for system-wide change efforts (Berman & McLaughlin, 1977; Deal *et al.*, 1975; Zaltman *et al.*, 1973). More surprising, however, is that within these 10 districts there seems to be a zero-sum relationship between the influence of various parties and the scope of change. A competitive tension may exist between hierarchical levels which has major implications for the process of initiating and implementing planned changes on a district-wide level. In discussing this possibility, a number of factors may help to illuminate the results, and will have bearing upon the general question of influence and changes in schools.

Teacher Autonomy

One question raised by our data is why teacher influence should have an apparent lack of impact. It appears that the sphere of teacher influence is primarily the classroom. Within our sample, for example, teacher influence in ongoing organizational decisions is positively correlated (.38) with teachers' assessments of the degree to which they are able to function autonomously in the classroom. The right to classroom autonomy is jealously guarded by highly professionalized teaching staffs, even in rural schools, and schools with more powerful teachers may tend to resist innovations that emerge from the superintendent's office.

Resistance may be particularly high in the case of innovations that attempt to standardize teaching procedures or curricula across classrooms, thus reducing the range of "professional judgment" available to the teacher. Resistance to district-wide changes on the part of influential teachers is not necessarily inconsistent with previous findings that staff professionalism is associated with innovations in schools. The important issue is the origin of the innovations proposed for implementation.

One of the frequently expressed goals of participation in planning is to reduce resistance to the change at lower organizational levels (Hage & Aiken, 1970). It is important to note, however, that within our sample, teacher and superintendent influence over planning are negatively correlated. Thus, as teacher input into the planning process increased, the critical superintendent role became weakened, and the *overall impact* of teacher participation apparently dwindled in its absence.

The Structure of Role Relationships in Schools

Schools within school districts are typically structured as segmented rather than cooperative (Bidwell, 1965; Deal *et al.*, 1975). In the segmented professional organization, each professional group has its own sphere of influence, and the activities of one may have little or limited impact on the activities of another. Increases in teacher autonomy in a traditionally structured school, for example, do not necessarily increase the influence of teachers as a group upon the activities of the school, but only the influence which each exercises in his or her protected domain. Under these conditions, increasing influence within the segmented units should not logically lead to more concerted joint effort at achieving a cooperative goal, since professionals in education are not accustomed to working under conditions where the development of common solutions to common problems is valued. The zero-sum interaction of influence found in our data substantiates the notion that under normal conditions schools are segmented not only physically, but also in terms of decision making.

These "normal" conditions of segmentation may be temporarily altered when the planning of a centralized change program—particularly one dependent on external funding—is at stake. Because this is an unusual event, principals and teachers may form a cooperative unit in an attempt to incorporate specific school needs and problems into a change program. If, as was generally the case in the Rural ES program, such cooperative planning efforts are limited to an initial planning phase, the cooperation is a temporary phenomenon, and the more typical patterns of segmentation can reemerge during the implemenation phase. While some substantial effort to develop strategies for ongoing cooperation might counteract this tendency, in the ES program such efforts (e.g., school or district-wide training programs) were generally limited to the *initial* implementation phases—an indication that participatory planning processes were not followed up by participatory management of the implementation effort.

Principal Leadership

The limited impact of both types of principal influence on implementation was one of the more surprising findings, since previous research has indicated that strong principal leadership is positively related to change (Berman & Pauly, 1975; Forehand, Ragosta, & Rock, 1976). The segmentation of schools provides only a limited explanation for the apparent lack of impact of principal influence. Principals occupy a position of middle management in the school district which may be characterized by stress in a period of district-wide planned change. On the one hand, their professional staffs expect them to represent and protect professional interests in the classroom, and to moderate what may be thought of as unreasonable incursions by the central office upon professional authority. Their supervisors, on the other hand, may expect principals to act as administrative representatives of the central office in ensuring the implementaion of its innovative initiatives. However, picturing the principal as a powerless functionary caught between opposing forces contradicts our data, which suggest that principals do have relatively high levels of general administrative influence within the school system.

One explanation for the apparently slight relationship between principal influence and the outcomes of the change programs may be found if the three measures of implementation are examined in greater detail. It will be recalled that the total scope of change was composed of two dimensions—quality and quantity. If we examine the regression of all authority variables on either quantity or quality, we find that principal authority has little impact. However, if we add an interaction term computed by multiplying the principal's influence, we find that this term has a strong positive relationship to quality of implementation (Table 27). (The same interaction term has no effect on quantity.) We also find that the introduction of the principal interaction term does not affect either the strong positive relationship between superintendent authority and implementation or the negligible one between teacher authority and implementation.

The quantity measure reflects indicators of change that are both highly visible and easily mandated by a superintendent. Quantitative

Table 27. Standardized (Beta) Coefficients for Regression of Quality of Change on Selected Authority Variables

Authority variable	Standardized regression coefficient
Main terms	
Planning influence	
Superintendent	.53[b]
Principal	–.03
Teacher	.03
Decision-making authority	
Superintendent	–.04
Principal	.10
Teacher	–.09
Interaction term	
Principal planning influence–principal decision-making authority	.30[a]
Multiple R^2—main terms	.28
Adjusted multiple R^2—main terms	.17
Multiple R^2—with interaction	.35
Adjusted multiple R^2—with interaction	.23

[a]*F* statistic significant at the .10 level.
[b]*F* statistic significant at the .05 level.

change can be achieved in the absence of leadership and commitment to innovation, for many people can be involved for substantial periods of time, but the content of activities can be essentially the same as that which existed previously. Ensuring that the change actually represents a difference in the structuring, content, and management of the school, however, represents a more dramatic and enduring aspect of innovation, one which the quality dimension attempts to capture. Interpreted in this light, the data suggest that achieving real changes in the operations of the school requires school-based leadership from the principal, in addition to that provided by the chief district administrator.

It is important to reiterate that it is the segmented nature of the educational system, and of the authority structure within the school system, that is most likely the crucial factor in explaining the outcomes of this analysis. Until the American educational system reflects structural changes that increase the levels of cooperation, integration, and participation in the ongoing functioning of education, the implementation of the type of *comprehensive, system-wide* changes intended by the ES program may be feasible only in a small number of districts with unique characteristics. Few would be satisfied with the simple recommendation that to facilitate change it is necessary to reduce the teacher role to a nonprofessional and subordinate status. If the goal is to attain a system-wide, innovative program, it will first be necessary to build a tradition of cooperative activity within schools and within the district as a whole which can serve to reduce the segmental nature of the authority structure while at the same time not undermining the leadership contributions available from professionals at all levels within the system.

The ways in which leadership, inappropriately exercised in a noncooperative context, can inhibit innovation may be seen in a number of examples in the ES districts. Butte–Angels Camp illustrates how a strong authority or leadership role can be inappropriate or counterproductive in a district where teachers have become accustomed to a great degree of automony in their classrooms. In that instance, teachers' recommendations solicited during the planning phase represented a wide variety of curricular innovations, many of which would have allowed for a more individualized approach to classroom instruction. The project plan, however, written by the superintendent and his select team, focused on the narrower approach of diagnostic prescription and instruction, to be applied uniformly to all aspects of the *existing* curricula. Thus, in addition to having had their expectations unfairly raised, teachers were burdened with additional paper work which they reported as having no beneficial effects on their teaching approaches and overall classroom routines. It was not until authority was somewhat decentralized during a major restructuring of project governance (which was designed to focus on teachers' problems with individual or small groups of students and to deemphasize teacher documentation) that teacher resistance to the project abated and new innovations were attempted (Firestone, 1980).

Clayville's experiences were similar in many respects. The superintendent was accustomed to planning *for* people rather than *with* them, and, like the superintendent in Butte–Angels Camp, presented teachers with new administrative and instructional burdens. Although he was a persuasive and powerful leader, who had demonstrated his ability to rapidly transform the schools so that they qualified for state and regional accreditation, he antagonized teachers and was therefore unable to elicit their cooperation in adopting the kinds of attitudinal changes that would lead to successful implementation of the district's ES project. Had it not been for his influence with the school board and the fact that teachers were afraid of being unable to find jobs elsewhere, it is unlikely that teachers would have cooperated to the extent they did (Clinton, 1979).

The second project director in Desert Basin was, on the other hand, a strong leader whose personalized, charismatic style engendered fairly widespread support for project activities despite the considerable demands he placed on staff members. His strategy of personally hiring and cultivating his staff, rather than creating a climate of apprehension over outside intrusion, was taken as evidence of his enthusiasm and commitment to the project, thus fostering cooperation among a rather large administrative group which generally filtered down to the teaching staff. Even such a widely acclaimed style of leadership has its hidden pitfalls, however, as was borne out by the demise of the project upon the departure of its leaders, whereupon it was left largely without support or new sponsorship (A. Burns, 1978).

Liberty Notch provides a striking contrast to the team efforts of the Desert Basin project. Not only was Liberty Notch plagued by an ineffectual first project director who failed to work cooperatively with his planning committees—to the extent that programs were not ready for implementation on time and had to be designed at the beginning of the so-called implementation phase—but there was little evidence of administrative support for the project throughout the union. This lack was largely attributable to the fact that the union was comprised of three autonomous school districts with no history of collaboration of the type required by ES/Washington. Consequently, it was not until building principals were made accountable for implementation of the project's components in their individual buildings that the project really got under way. Throughout the life of the project, the retention of the first project director as career/vocational coordinator rankled, since many staff members attributed lack of cohesion and momentum to his poor leadership (Stannard, 1978).

Such examples signal the need to interpret with caution the notion of leadership and to offer the caveat that, while forceful and concentrated (strong) leadership may be necessary for the successful implementation of planned change, it is essential that the leader is, in some sense, a member of the group from which he derives his support and compliance. Someone who is viewed by the implementing group as the "enemy," or, at best, as a disinterested outsider, cannot effectively deal with the problems and concerns of the staff. Furthermore, it is necessary to consider various factors and perspectives when evaluating the success of an implementation. In both Butte–Angels Camp and Clayville, the project leaders were relatively successful at least for a while, in enforcing compliance—which could be pointed to as successful implementation of their objectives. Teachers and pupils, however, would be likely to argue that documentary evidence of implementation, such as the completion of diagnostic instruction sheets for a specified percentage of students, does not provide an accurate reflection of the degree or kind of change taking place in actual classroom learning.

HOW DOES SYSTEM LINKAGE AFFECT SCHOOL IMPLEMENTATION?

The impact of authority variables on the implementation of the ES program directly addresses only one facet of the system linkage, as we defined it in Chapter 2. Among the structure and culture variables, there are six which have been identified as measures of the degree to which the schools in the Rural ES program correspond to several of the definitions of linkage.

Among the structure measures that may be considered indicators of system linkage are those dealing with the authority structure (among which we have selected superintendent authority, principal authority, and teacher autonomy) and the level of formalization. The authority structure variables are assumed to be related to system linkage through the degree to which they increase or decrease the *coordination* among units of a school or district. Where potential administrator influence over decisions is high, we assume that coordination will be higher. Where teacher autonomy within the classroom is high, on the other hand, we assume that coordination among units, both directly at the school level and secondarily between schools, will be more uncertain.

Formalization is an important measure of system linkage because the making and enforcing of rules is one of the major ways in which bureaucratic organizations ensure that there will be some standard operating procedures to govern behavior across units. In our measure of formalization we have taken account of the fact that many schools are "mock bureaucracies" (Gouldner, 1959) that have many rules, few of which are enforced, and thus we measured not only rules, but *enforced* rules.

Cultural linkage measures the degree to which the climate of the organization reflects interaction between individuals. We assume that the greater the amount of work-related contract among individuals, the greater the cultural linkage. Although we had available to us no direct measures of contact, two of our variables seem to be good proxies for cultural linkage. The first is collegiality, which is composed of measures indicating the peer supportiveness of the staff. The second is the level of disputes, which we interpret as a positive indicator of linkage since the level and frequency of disputes among units that do not interact with one another is likely to be low. The fact that collegiality and level of disputes are negatively correlated ($r = .51$) does not undermine this argument, for we assume, like Weick (1976), that it is not necessary for all indicators of linkage to be consistent, nor is it likely that all possible arenas of "tight coupling" will occur simultaneously. In fact, it is possible that increased communication between role partners within a school may lead to either increased collegiality or increased conflict, depending on such other variables as the nature of the systems and/or the issues involved.

Another measure of linkage that is not reflected in our previous discussions is the degree to which the school is physically proximate to the other units within the district. Given the rural context of the ES program, the concept of physical linkage is perhaps one of the most powerful issues. We assume that physical proximity increases the likelihood of communication among units and that greater distances increase problems with both communication and other kinds of coordination (R. Hall, 1972; Louis & Sieber, 1979).

Three measures of a school's physical linkage with other units in the district were used in the study: the distance to the nearest school, the distance to the farthest school, and the distance to the district central office. In several of the districts (Salmon Point, Desert Basin, and Prairie Mills), all or most of the schools are located on a single campus or within the same village. In other districts, the schools are widely dispersed, Big Sky being the most extreme case.

Not suprisingly, the three members of physical linkage are quite highly related to one another. (Pearson correlations ranged from .41 to .76.) Therefore, the measures were standardized and summed into a single index of physical isolation. The simple correlations between this measure and total scope ($r = -.44$) and between this measure and quality ($r = -.55$) indicate a strong negative relationship.

Two issues regarding system linkage and implementation at the school level have now emerged. The first is the power of the structural- and cultural-linkage variables to predict implementation in the absence of other powerful predictors (such as size, morale, and staff educational level). The second is the degree to which physical-linkage problems, exacerbated by the rurality of these districts, affected implementation.

Structural and Cultural Linkage as Predictors of Implementation

Table 28 presents the results of a multiple regression of total scope, quantity, and quality on the linkage variables, and on interactions between pairs of linkage variables. The multiple R^2s at the bottom of

Table 28. Standardized (Beta) Coefficients for Main and Interaction Terms in the Regression of Each of Three Measures of Implementation on Selected Linkage Variables (N = 45)

Linkage variable	Implementation measure		
	Total scope	Quality	Quantity
Main terms			
Collegiality	.46[b]	.09	.52[b]
Disputes	.27[a]	.15	.22
Formalization	−.06	.13	.01
Superintendent authority	.37[b]	.51[b]	.25[a]
Principal authority	−.18	−.09	−.18
Autonomy	−.42[b]	−.19	−.49[b]
Interaction terms			
Disputes/Formalization	−.31[b]	—	−.35[b]
Autonomy/Principal authority	—	—	−.24[a]
Collegiality/Principal authority	—	.36[b]	—
Multiple R^2			
Main terms	.45	.37	.40
Main plus interaction terms	.51	.45	.52
Adjusted multiple R^2			
Main terms	.36	.27	.31
Main plus interaction terms	.42	.35	.41

[a]*F* statistic significant at the .10 level.
[b]*F* statistic significant at the .05 level.

Note: For each of the three regressions, all of the main terms were entered prior to any of the interaction terms. Beta coefficients for the interaction terms are presented for only those variables that would increase the stepwise multiple R^2 by at least 2%.

the table indicate that on the whole, linkage variables are almost as strong in predicting total scope or quantity as are a combination of the most powerful nonlinkage predictors from the structure, culture, and input groups (see Table 29). In the case of the quality dimension, however, the difference is only 1% of the variance. If we include a single, powerful interaction term (that between principal authority and collegiality) in the equation with the main-term linkage variables, the multiple R^2 is far higher.

What does this finding suggest? First, it indicates that variability in the degree to which schools exhibit characteristics of loose or tight linkage among parts does, in fact, have a significant effect on system behavior, and that this system effect is as important as other powerful structure and culture features. This finding adds considerable credibility to the contention that the concept of loose system linkage may have significant theoretical contributions to offer for an understanding of

Table 29. Standardized (Beta) Coefficients for the Regression of Each of Three Measures of Implementation on Selected Nonlinkage Variables (N = 45)

Nonlinkage variable	Implementation measure		
	Total scope	Quality	Quantity
School size	.22 (3)	.30[a] (3)	.12 (4)
Individualization	.32[a] (2)	.22 (4)	.34[a] (2)
Morale	.06 (6)	.24 (5)	—
Problems	−.07 (5)	−.06 (6)	−.06 (5)
Educational level	.49 (2)	.34[a] (1)	.51[a] (1)
Percentage male	−.18 (4)	−.14 (2)	−.17 (3)
Multiple R^2	.48	.38	.44
Adjusted multiple R^2	.40	.28	.37

[a]Nonstandardized regression coefficient is at least twice its standard error.

Note: For each of the three regressions, beta coefficients are presented for only those variables that would increase the stepwise multiple R^2 by at least 2%. The sequence of variable entry into the regression equation was unforced. (Order of entry is presented in parentheses below the associated coefficient.)

organizational behavior. But it also indicates that "loose coupling" is an inadequate and incomplete explanation of the variability in the ways that systems behave. While March and Olsen (1976) may be correct in terming organizational decision making as a "garbage can" process, in which all sorts of random factors get mixed together until a decision is expelled, in this instance the outcomes of one form of organizational decision making (i.e., implementation of planned comprehensive change) appear to be less random, and strongly related to the degree to which the system resembles organized (as opposed to disorganized) chaos.

A second conclusion one may draw is that, while the quality of organizational change may be more difficult to predict than its quantity, linkage variables make a relatively more powerful contribution to the explanation of qualitative change. In other words, disconnected or loosely linked systems may be capable of instituting changes which require some simple modification of existing practices, even if the modifications affect all units within the system. Where a more substantial change in existing procedures or behaviors is required, however, the loosely linked system is relatively less adaptable, at least when the innovation is expected to affect several parts of the system simultaneously.

Physical Linkage: The Rural Factor

The ES program was designed to draw upon the strengths of rural communities to build programs that took rural characteristics into account. The degree to which the isolation or rurality of schools actually affected the implementation of the ES program is therefore of interest. Our examination of this point was conducted with regressions of total scope, quantity, and quality on selected powerful variables from the structure, culture, and input groups, including in each regression equation the new variable of physical isolation. The results of these regressions are presented in Table 30.

These data reveal that, on the whole, the physical isolation of any given school was not the most significant factor affecting implementation of its ES project. When we examine the regressions on total scope, it is clear that other variables—superintendent authority, school size, level of individualization, morale, collegiality, and all three input

Table 30. Standardized (Beta) Coefficients for the Regression of Each of Three Measures of Implementation on Structure Variables and Isolation, Culture Variables and Isolation, and Input Variables and Isolation (N = 45)

Variable	Implementation measure		
	Total scope	Quality	Quantity
Structure and isolation			
Superintendent authority	.38[b] (1)	.38[b] (1)	.31[b] (1)
Autonomy	−.17 (2)	−.16 (3)	−.16 (4)
Size	.29[b] (3)	.28[b] (2)	.26[a] (3)
Individualization	.27[b] (4)	13 (4)	32[b] (2)
Isolation	.02 (5)	−.08 (5)	.09 (5)
Multiple R^2	.40	.41	.31
Adjusted multiple R^2	.32	.33	.22
Culture and isolation			
Morale	.34[b] (2)	.43[b] (3)	.23 (4)
Collegiality	.67[b] (1)	.40[b] (2)	.76[b] (1)
Problems	.15 (5)	.07 (5)	.18 (5)
Tension	.06 (4)	−.18 (4)	.22 (2)
Isolation	−.28[b] (3)	−.40[b] (1)	−.15 (3)
Multiple R^2	.41	.37	.39
Adjusted multiple R^2	.34	.29	.31
Input and isolation			
Percentage Male	−.37[b] (2)	−.27[b] (3)	.31[b] (2)
Educational level	.51[b] (1)	.38[b] (1)	.52[b] (1)
Modernity	.25[b] (3)	.16 (4)	.27[b] (3)
Isolation	−.17 (4)	−.31[b] (2)	−.05 (4)
Multiple R^2	.47	.37	.43
Adjusted multiple R^2	.42	.30	.37

[a] *F* statistic significant at the .10 level.
[b] *F* statistic significant at the .05 level.
Note: Order of entry is presented in parentheses below the associated coefficient.

variables—far outweighed the importance of isolation, despite the high simple correlation of this rural linkage indicator with total scope. The same finding also applies to quantity of implementation.

Turning to quality, however, isolation appears to be considerably more significant. In equations with culture and input variables, isolation emerges as the first or second most important predictor of implementation. Only in conjunction with the structure variables does it have little impact upon the quality of change.

The fact that isolation has an independent impact upon the quality of implementation is surprising, considering how highly related it is to other factors that are associated with change. Isolation, for example, is positively associated with morale ($r = .30$), but is negatively associated with other input and culture variables that are postive predictors of change: educational level of staff members ($r = -.37$), tension ($r = -.31$), perception of educational problems in the school ($r = -.30$). The concept of physical isolation (loose linkage) may be not only a surrogate indicator for other school characteristics which apparently impede change, but may also make an independent contribution.

The way in which physical isolation is related to the structure of a school demonstrates even stronger relationships. The more isolated the school, the more likely it is to have little perceived influence from the superintendent ($r = .41$). In other words, it is not only physically isolated, but also isolated from external influence. In addition, isolated schools tend to be quite small ($r = .51$). As we have seen time and time again in our analyses, in rural school systems the role of the administrative authority structure and the impact of school size are critical in predicting the degree to which schools adopt innovative programs of the type sponsored by the Rural ES program.

These data point strongly to the conclusion that even a relatively intensive program of support to rural schools that are located in physically isolated areas will not result in their using the curricular advances of the last 15 years or so. If the society values school innovativeness above those school characteristics that may be best represented by the traditional rural school—congruence with community values, local autonomy, congenial interpersonal realtionships—then our data provide support for school consolidation as a "solution." On the other hand, if the contemporary regard for innovativeness in education is a value based on ideological preferences rather than proven effectiveness, the traditional rural values may be no less valid.

CONCLUSIONS

At the beginning of this chapter it was indicated that our analysis would address three enduring issues in the field of organizational and educational change. The first of these concerned an examination of alternative theories of change that are based on sociological (structural), social-psychological (cultural), or staffing (input) variables. A second concerned the importance of control, authority, and participation in the facilitation of organizational change. The final issue addressed was the relative importance of system linkage as a factor in explaining change outcomes. Since this chapter has been relatively long and complex, we will attempt to draw some conclusions about the way in which our study has illuminated each of these issues.

Alternative Theories of Change

The facilitation of organizationwide innovative programs requires many sources of organizational support. Structural support, a suitable organizational culture or climate, and a staff that exhibits certain important group characteristics are all equally important to the change effort. An integration of alternative theories of change does not, however, involve simply adding various measures to a simple model of a "best" organizational environment for change. Rather, our examination of interaction terms indicates that there are strong contingent effects in operation. Among the most important findings in this analysis are the ways in which cultural variables and staffing variables moderate the often-replicated association between organizational size and program innovation, and the ways in which powerful organizational control effects on change are influenced by staff cultures. The importance of these interaction terms lies not simply in helping us to better understand the meaning of our data on the change process and outcomes in rural schools. Rather, it has the additional significance of pointing to the need to move away from simple linear models of the change process toward more complicated contingency models. The use of interaction approaches in the development of contingency theory can assist in delimiting the number of contingency situations that are to be dealt with—a limitation that is essential if contingency approaches are to be of theoretical and practical utility. Although we would interpret the strong interaction or contingency effects associated with the variable "percentage male" to be uniquely attributable to the rural context of our study, the importance of size, control structures, and the use of individualized (as opposed to standardized) technologies should be further explored in the development of contingency-theory approaches.

Participation, Authority, and Implementation of Organizational Change

The finding that strong, centralized authority facilitates program change, and that participation—particularly participation in the planning—may impede it, is clearly a controversial one, for it undermines many of the premises associated with human relations approaches to management. Of particular importance, in our opinion, is the finding that in these rural schools, patterns of influence interact in a zero-sum fashion. Participatory theory would argue that when the total amount of influence in an organization is increased, the effectiveness of the organization is also likely to increase. We find, however, that this is the case only with regard to the interaction between principal and teacher influence on decision making. All interactions between school-based personnel and the superintendent indicate that increasing the amount of influence within the system through participation decreases effectiveness (in this case, implementation of comprehensive change).

We feel that it is important to interpret our findings with caution. In particular, the finding that strong, centralized leadership facilitates change should not be interpreted to mean that the innovative programs should be thrust upon protesting organizational members with imperial disregard for their concerns. Instead, we need to examine the entire context of our findings, and some of the reasons why participation may fail to improve the chances for successful implementation. Among those that we have suggested are the lack of collaborative experience, particularly in rural and/or traditionally structured schools, and the program-specific requirement that all changes be made comprehensively across several school units. In addition, it is necessary to add the caveat that our data do not allow us to examine in detail how the impacts of centralized leadership may have been played out over the years *after* early implementation. What was covert teacher resentment of the program during implementation may, over the course of time, have resulted in more extensive discontinuation and disaffection as superintendents turned their attention to new tasks and programs. Despite these cautions, however, it is clear that we must reexamine the meaning of broad-based participation in program innovation, and determine approaches to planning and implementation that takes findings such as those presented above into account.

Of particular importance, in terms of our theme of organizational linkage, is the finding that principal authority makes little difference in the case of quantity of change—a measure that is a highly visible, easily enforceable, external indicator of implementation. In the area of quality of change, or the degree to which the change actually represents a difference in practice, influence from the school-based administration becomes much more significant. This finding highlights the fragility of linkages between school and district, and some of the reasons for persistent findings in other studies (Cohen, Deal, Meyer, & Scott, 1976; Deal *et al.*, 1975) that district-level variables have insignificant relationships to school-level behaviors.

Linkage and Implementation

Measures of linkage—culture linkage with schools; structural linkage within schools, between school and district, and in the physical isolation of schools—appear to be significant predictors of organizational behavior in schools. While other scholars have implied that linkage may affect such behavior, previous empirical studies have not moved beyond searching for proof of the existence of variable linkage in systems. We, however, have started from a position of assuming variability in the "tightness" or "looseness" of linkages, and have attempted to show how this variability will affect the propensity of organizations to adopt system-wide changes.

It is important to emphasize that the findings should not be interpreted to suggest that tightly linked systems are necessarily more innovative. As Weick (1976) points out, one of the advantages of loose coupling is that it allows adaptations in different parts of the system, or localized "invention." We have not examined this phenomenon at all, and it may well be that those schools low in system-wide implementation of the type intended by the ES program may still be engaged in considerable local adaptation.

Such a form of educational innovation, however, is not only

"piecemeal," as defined by the originators of the Rural ES program, but also frequently low in impact because it is limited to single classrooms in a school, or single schools within a district. However, as we shall show in the next chapter, piecemeal change is not necessarily little change; nevertheless, isolated change that does not have any impact beyond the initial locus of a spontaneous innovation does not necessarily further the social goals of education in the short run. The concern with equity in education today demands that attention be paid to raising the less innovative schools to a minimal level of comparability with the more innovative ones. Our data indicate that the spread of an innovation within a given system is less likely if the system's parts are only loosely linked.

8

Comprehensive Change at the District Level

The analysis of implementation at the school level, although important for understanding the process of planned educational change, should not obscure the fact that the Rural ES program was designed as a district-wide phenomenon. Neither entry into the program nor planning the ES projects was a school-based activity; rather, they were centralized in the school district office. The 45 schools whose patterns of implementation were analyzed in the two previous chapters were all participants in the Rural ES program, not as individual schools, but as units of school systems. Thus, they were embedded in 10 discrete districts, each of which undertook to implement a district-wide plan for change. And, although there was considerable variation in implementation among schools within a given district (see Figure 4, p. 134), such variation was clearly greater in some districts than in others.

This chapter and the one which follows examine differences in implementation at the district level with the objective of understanding why some districts were more successful than others in implementing a program of planned change. We first present a measure of implementation that takes into account special difficulties involved in measuring district-wide, as opposed to school-based, change. We then address the question: Did the ES districts implement truly comprehensive change, or did they stress some facets of comprehensiveness more than others?

Using our measure of district-wide implementation, we go on to examine the relationship of two types of factors with level of implementation. The first type includes factors associated with the design of the ES projects themselves, including ways in which the projects were structured, the amount of funds allocated to planning, and the strategies which were used to implement the projects. A second set of factors concerns two features of the organizational context in which the projects were implemented—the level of support of local staff and parents for the ES projects, and the nature of the administrators' role in the district. Of particular interest is the effect on the level of implementation of different types of linkages (structural and normative) between teachers and administrators. (Additional organizational factors and their effect on implementation will be discussed in Chapter 9.)

MEASURING IMPLEMENTATION AT THE DISTRICT LEVEL

In order to answer questions about the degree to which the districts were able to implement comprehensive change, it was necessary to construct a measure of implementation at the district level. The scope-of-implementation score described in Chapter 6 served as a framework for examining ES implementation in schools. One way to determine district-level implementation would be to take the *average scope score of the schools* in a district as a measure of district "success." Though this approach may seem generally adequate, it would ignore one of the main features of "comprehensiveness," which was the district-wide nature of the ES program, and the notion that the whole was to be greater than the sum of its parts. Furthermore, each district designed project components that were not intended to operate at the school level, such as community participation, adult education programs, and improved administrative and evaluation techniques, to name only a few.

As a consequence, in order to capture the "whole" of ES at the district level, it was necessary to construct a score that reflected the basic unit of project planning and innovation: *the project component.* Here, however, we faced one of the serious problems in measuring organizational "innovativeness" (see Chapter 2): how to compare "apples and oranges." In concrete terms, each district interpreted the notion of project components very differently; indeed, the nature of components could vary greatly even within districts. A component could be as discrete as the adoption of a specific social science curriculum for a single grade level, or as broad as the adoption of diagnostic instruction across all grade levels and subject matter areas. We needed to measure implementation in a way that took into account these varying conditions.

It will be recalled that, in Chapter 3, we presented a matrix that we designed to measure the implementation of educational innovations (see Table 3, p. 64). The first step in developing a district-level scope was to find indicators for each cell of this matrix for each component in each district (Table 31). The requisite data were obtained from questionnaires completed by the on-site researchers in the spring of 1975, two years after the beginning of the implementation phase, thus permitting generation of dimension and facet scores and a total scope-of-implementation score for each of the 129 components that were "active" at that time.[1]

A second major step in computing a district's scope-of-implementation score was to aggregate the scores for each component within a district's project to obtain that district's implementation score. But such an aggregate score posed some conceptual and methodological problems, largely because the districts varied so widely in the *number of components* that they chose to plan and implement. Districts such as Big Sky, Magnolia, Desert Basin, and Salmon Point designed programs with fewer than 10 components. At the opposite extreme was Timber River, whose ES project resulted in the implementation of

[1]The district-level quantity-of-change dimension score was computed by summing each of the columns for the subdimensions of pervasiveness and extent and then adding them together. Quality of change was computed by summing the 14 indicators listed in Table 31. The indicators of quality of change for the time, space, and facilities cell were weighted in order to ensure that each facet contributed equally to the final score.

In computing district-level facet scores, equal weight was given to the two dimensions of quantity and quality, and individual indicators were weighted to accomplish this end. The items were then added across rows to produce the facet scores. Both dimensions and facets were normed to range from 0 to 100.

The district-level total scope-of-implementation score for each component, the final product of these calculations, represents the addition of the marginal score (either dimensions or facets), again normed to range from 0 to 100. The internal reliability of this summary measure is .74 (Cronbach's Alpha). A copy of the instrument used is found in Appendix B.

Table 31. Indicators for Scope of Change Matrix for Each Project Component

Facets of educational change	Dimensions of educational change			Overall facet scores
	Quantity of change		Quality of change	
	Pervasiveness	Extent		
Curriculum	(1, 1) Percentage of all pupils involved	(1, 2) Average percentage of pupil time	(1, 3) How different curriculum is now; Number of new techniques; Amount of change in classroom organization	*F*1
Staffing	(2, 1) Percentage of all teachers involved	(2, 2) Average percentage of teacher time	(2, 3) How different staffing is now; Number of new techniques; Amount of change in interpersonal skills for staff	*F*2
Time, space, facilities	(3, 1) Degree to which these three entities used in a changed way, expressed as a percentage	(3, 2) Percentage of time, space, and and facilities used in a changed way in addition to or in comparison to previous uses	(3, 3) How different is use of time, space, and facilities; Number of new techniques in time, space, and facilities	*F*3

Community participation	(4, 1) Sum of parents, preschoolers, senior citizens, and others involved divided by total number of students and expressed as a percentage	(4, 2) Average of average percentages of time for 4 groups in cell to left	(4, 3) How different community participation is now Number of new techniques used in community participation Form of community participation	*F*4
Administration, organization	(5, 1) Percentage of all administrative personnel with major responsibility	(5, 2) Total percentage of administrative time divided by largest possible number of administrators	(5, 3) How different is administration and organization Number of new techniques in administration and organization Amount of influence on change for various groups	*F*5
Overall dimension scores	*d*1 *D*1	*d*2	*D*2	Total scope score

40 components by 1974–1975. The remaining districts ranged from 11 to 17 components.

Since federal project funding and the available local resources of time and personnel were not substantially greater in those districts with a large number of components than in those with only a few, it seemed that the districts had chosen *different strategies* in designing their ES projects. In order to achieve the same district-wide level of comprehensive change, a district with a small number of components would have to make each component broader than a district with a larger number of components. In the latter case, comprehensive change might be achieved by an accumulation of small innovations involving multiple programs targeted at more specific objectives.

With such variation in the number of components, it did not seem reasonable to assume that a score for any district computed by simply averaging across its components would accurately reflect the degree to which the district had implemented changes. Two districts might, for example, have roughly the same average-component score, but one district might have twice as many components as the other. Clearly the existence of more components in such a case suggests a different form of change which needs to be captured in any district-level scoring process.

Although it is important to avoid using an average-component score, simply adding together the separate component scores for each district would introduce another type of bias. Any component, no matter how small in its intended and actual scope, would be weighted equally. The problem of potentially overestimating a component's effect within the district is minor for the quantity dimension, for here we are looking at actual percentage counts of people and time, which are comparable quantities both within a single district, and across districts. In the measurement of quality, however, we were interested in *how different* the new component was from what existed previously. Here it is conceivable that a component that was relatively narrow might still represent quite a major difference from previous practices, yet have a small impact within the district as a whole.

In aggregating across components, we chose to compromise between using average and additive scores by weighting the average score by a factor which reflects the number of components—namely its square root. The use of the square root of the number of components, as compared to using an additive model, also has the advantage of reducing the range of differences among the districts.[2]

Two possible measures of implementation at the district level are presented in Figure 5A: one is the average of the *school* scope scores in each district, and the other the weighted *aggregate of component* scope scores in each district. A comparison of the two shows that the relative position of the districts remains remarkably stable, no matter which measure is applied.[3] Since each measure taps a part of the program outcomes that is ignored by the other, this essential stability is particularly reassuring. Using either method of scoring district-level implementation, Timber River implemented the most, followed by Salmon Point, Prairie Mills, and Oyster Cove; relatively little was implemented in Big Sky, Butte–Angels Camp, Liberty Notch, or Desert Basin. The only major discrepancy occurs in Clayville, which has the second-highest score when *school* scores are averaged, but scores lower when aggregate *component* scores are used. (The probable reason for this discrepancy is the early discontinuation of Clayville's Diagnostic Reading Center, a major, nonschool-based innovation which was abandoned in favor of locating similar [although minor] activities within schools.)

Certain other differences between the total scores of school districts are also apparent in Figure 5A. The range of scores is somewhat greater for the aggregate *component* score (from a high of 92 to a low of 33) than for the average *school* score (a range of 72 to 28). In terms of aggregate component scores, school districts score either high (above 70) or low (below 50). For the average school scores the distribution is somewhat more continuous.

We took a third and final step in computing a district-level scope score which could be used in the analysis of factors related to district implementation. A total scope-of-implementation score was computed by taking the *average* of the aggregate component score and the average school score for each district. The decision to proceed in this

[2]The reduction in the spread of the scope scores across the 10 districts corresponds to the interpretive judgments of the OSRs, whose ethnographic data revealed differences in implementation among the districts that correspond to the quantitative data (see Herriott & Gross, 1979, Chapters 4–8).

[3]The Spearman's Rho for the 10 school-district scores computed by the two methods is .75, which is significant at the .05 level.

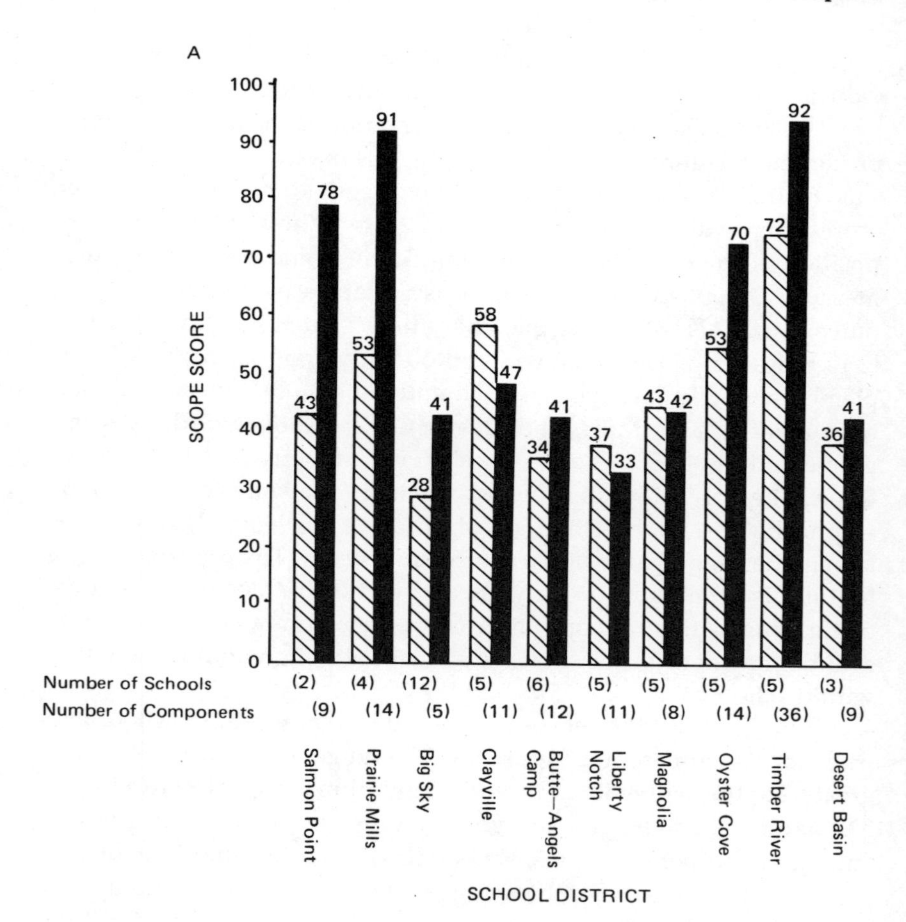

Figure 5. Total scope of implementation at the school-district level: developing a summary measure. (A) Two measures contributing to the total school-district scope score. ▧ = average of *school* scope scores in each district; ■ = weighted aggregate of *component* scores in each district. (B) composite total school-district scope score.

way was based on the strong relationship between the two measures and the fact that each reflected a slightly different aspect of the implementation outcome. This *composite district score* is presented in Figure 5B. Before presenting our analyses based on this score, however, we need to consider the utility of the constituent measures of the aggregate component scores, measures of the individual facets of comprehensive change.

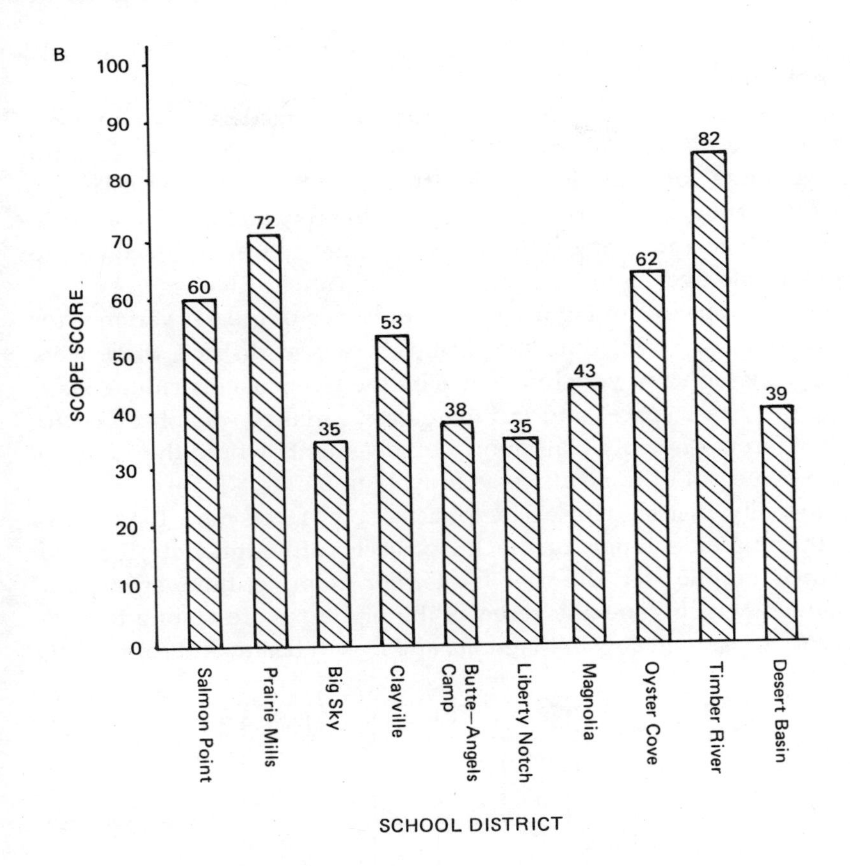

HOW COMPREHENSIVE WAS IMPLEMENTATION?

Comprehensive change, as originally defined by ES/Washington and as reflected in the plans of each school district, was to affect each of five facets of the districts' functioning: curriculum; staffing; time, space, and facilities; community participation; and administration and organization. However, in actual practice, the school districts varied in the strategies they chose for implementation and in the targets they chose to emphasize. Since the matrix presented in Table 31 was constructed to allow us to generate facet scores (*F*1, *F*2, *F*3, *F*4, and *F*5) for each component, we explored the degree to which the 10 districts achieved the objective of affecting each of the five facets of comprehensive change. To do this, we aggregated the facet scores across all components in a given district using the weighted procedure already described (Figure 6).

It is clear from Figure 6 that there are considerable differences among the districts in their scores for any given facet. Curriculum scores range, for example, from less than 35 in Big Sky to 150 in Timber River; staffing scores are as low as 22 in Big Sky and as high as 82 in Prairie Mills; and changes in the use of time, space, and facilities go from a little over 10 in Magnolia to 160 in Timber River.

Also apparent in Figure 6 is the presence of greater variance for some facets than for others. Specifically, across districts there were both very high and very low scores on the facets of curriculum, staffing, and time, space, and facilities. Less variance was observed for other facets—notably community involvement, where the range of scores in eight of the districts is from 17 to 50.

Finally, districts chose to emphasize some facets more than others in the implementation of the ES projects, although not all school districts emphasized the *same* facets. For example, time, space, and facilities, which generally receives the highest score among the five facets, is relatively low in Magnolia and Desert Basin, and community

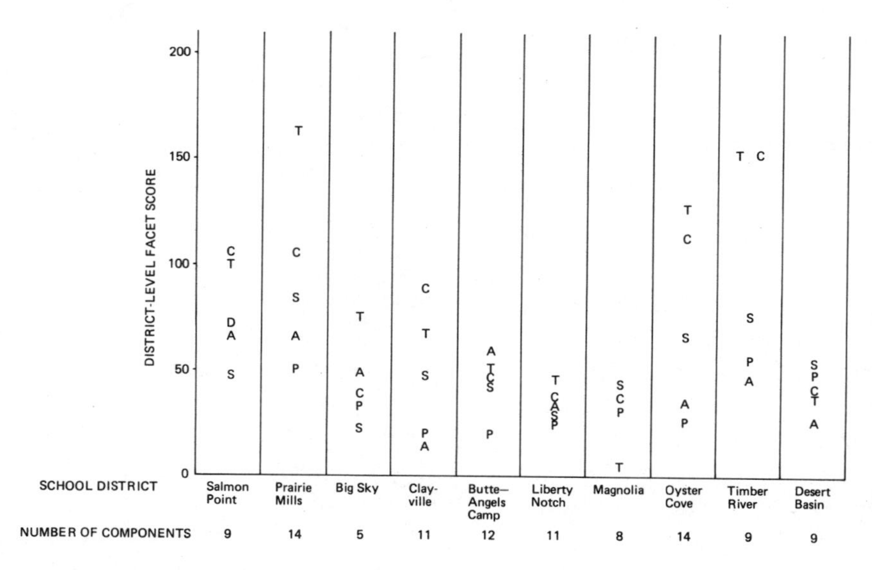

Figure 6. District-level facet scores, by school district. C = curriculum; S = staffing; T = time, space, and facilities; P = community participation; and A = administration and organization.

participation, which tended to score lower than curriculum, scored higher than curriculum in Desert Basin.

There is a strong tendency for *both* curriculum and time, space, and facilities to be high in sites with overall high implementation, suggesting that well-implemented curricular changes involve extensive revisions in the use of time, space, and facilities. We may further infer that during planning for new curricula, explicit attention needs to be given to how the use of time, space, and facilities may require alteration to be compatible with new curriculum objectives. This suggests that an "architectural approach" to making curriculum changes may aid in the implementation process.

Those sites which display high curriculum as well as time, space, and facilities scores show somewhat lower scores for staffing. Our initial expectation was that curriculum and staff training would go hand in hand and, in fact, the innovative focus within schools is typically upon the curriculum–teaching skills package. However, staff training in the Rural ES districts tended to be a "one-shot" activity rather than a continuous process. Many districts used outside consultants to design training sessions at the time of initial implementation, often because ES/Washington advised or required them to. But consultants were located too far away from the districts to be available on a continuing basis, and the districts were usually unfamiliar with and distrustful of the consulting process.[4] Such lack of integration between ongoing staff training and curriculum implementation created problems for sustained implementation and coninuation of curriculum-related activities. (See also Berman & McLaughlin, 1978.)

The school districts that exhibited higher scores in the staffing facet were those in which staffing was a separate and highlighted project component—that is, not connected with curriculum components. The highest score was received by Prairie Mills, whose "Improved Staffing" component received a proportionally high amount of funding during both the first and second years of the project. Timber River, with the next highest score, had a component called "Instructional Leaders" which designated a staff person in each building to serve as a resource person for the rest of the staff, and who was responsible for continuous maintenance of staff-training activities.

[4]See Lippitt (1979) for an extensive analysis of the use of consultants within the Rural ES program.

Thus, a continuous rather than a "one-shot" staff-training approach was deliberately adopted in Timber River.

Community participation, which was a priority of ES/Washington, was an ineffectually implemented component in most of the ES districts. The exception is Salmon Point, which had begun an active community coordinating council prior to the funding of ES. Even in Salmon Point, however, the level of participation was affected very little by ES project activities. Overall community participation was continued only as long as ES/Washington insisted on it (primarily in the planning year), and was then quietly allowed to disappear.

One explanation for the general lack of community participation is the tradition of conflict between educators and community members over educational matters. School administrators are frequently unprepared to share decision making with parent councils (Cohen *et al.*, 1976; Fantini, 1972). The Rural ES sites were especially unlikely to implement community participation since they generally had no established lay interest groups and little prior experience in active lay-involvement roles.

In some districts there was a discrepancy between what educational administrators viewed as good education and what parents viewed as good education. The superintendents, who had the greatest influence in planning and implementation, were for the most part trained in university centers and were more educationally sophisticated than their community constituencies. Either for that reason or because they sensed that individualization was a concept popular with ES/Washington, administrators tended to shape the plans in the direction of individualized education, whereas parents and community members tended to take a more traditional view of education—the "three R's" perspective. Given this difference in perspective, it would not be surprising if parents felt little enthusiasm about participating in ES project activities.

Administrators, in turn, recognizing that parents' goals differed from their own, and perhaps guarding their autonomy in educational matters, found ways to limit parent effectiveness. Lay people were usually not included in formal decision making. Community advisory committees, formed at the insistence of ES/Washington, were often cumbersome, met infrequently, and became forums for the administrators to inform the community about the ES project or other school activities rather than opportunities for citizens to design and propose changes in the school system. Lacking any real influence, committee meetings often turned into gripe sessions for parents. Community members began to see the meetings as rubber-stamp operations and lost interest in them. Teachers and administrators often viewed the meetings as trivial and a waste of time and very likely communicated this perception to parents. Many committees tapered off and then were discontinued during the implementation period; those that survived had little impact on the school systems.

Reorganization of administration and organization within the school district is clustered with community participation toward the bottom of the scale in all districts except in Magnolia, where it received the highest score, implying that major changes in lines of authority and administrative procedures are difficult to achieve in the context of a program such as ES.

It is also likely that those who were in authority were unwilling to tamper with the status quo and had little interest in change in this facet if it affected their own domains. Some of the districts had powerful superintendents who sought to extend their authority by means of their districts' participation in the ES program. The Desert Basin superintendent, for example, brought in coordinators to run the program who were viewed as the superintendent's "cadre of experts" and formed an overlay on the district's regular organization chart (A. Burns, 1978). The superintendent in Clayville hired the second ES coordinator on the basis that "he's loyal to me" (Clinton, 1979).

On the other hand, other districts had no tradition of highly centralized district leadership, particularly those that had recently consolidated (such as Big Sky and Butte–Angels Camp), and generally relegated authority to the building level. Regardless of the district administrative structure before ES, however, planned change altered it very little.[5]

The district that scored highest on this facet was, interestingly, Magnolia, the only district that did not really attempt to initiate district-wide programs. In this recently integrated Southern district, the ES effort focused on improving education within the formerly segregated schools. Many new administrative positions were created as a result of the ES project, which caused the district to score particu-

[5]Note that the absence of strong *planned changes* in administrative structure and procedure does not necessarily imply that there was overall stability in this facet, but only that these were not in the change arena *intended* by the ES projects.

larly well in the administration and organization facet. In addition, because of the recent consolidation, the task of managing a federally funded program introduced new central-office responsibilities that had not previously existed.

FACTORS RELATED TO IMPLEMENTATION

We return to our composite measure of implementation at the district level—the total school-district scope score presented in Figure 5B. Several sets of factors help to explain why some districts achieved relatively high scope-of-implementation scores while others did not. One set of explanatory variables focuses on the programmatic characteristics of the ES projects. Another set centers around the organizational and environmental characteristics of the projects. These latter factors were a major focus of Chapters 6 and 7 (for school-level implementation), and they will be discussed further in Chapter 9 in our summation of characteristics of districts which achieved relatively high levels of continuation. Only selected types of organizational characteristics—local support for the projects and the normative and structural linkages between role partners in the change process (teachers and administrators)—are discussed in this chapter in terms of their effects on implementation.

Programmatic Features Associated with Implementation

If we isolate the ES projects from their district contexts for a moment and examine only characteristics of the projects themselves, certain differences between the high and low implementers become apparent. Three of the four highest scoring projects—Prairie Mills, Oyster Cove, and Timber River—have a far greater number of components than any of the other projects, and these components are *highly differentiated*. In other words, the content of the components was usually narrowly targeted to a specific grade level, a narrow portion of the curriculum, or a highly specific effort to change parent/community involvement. This highly differentiated and targeted approach may be contrasted with that taken by Big Sky's project; it had only five components, which were very broadly defined.

One explanation for the positive outcomes associated with highly differentiated projects is that failures are easier to "bury" within such a large number of smaller scale components, and contamination due to failure in one area does not easily spread from one component to another. One of the lowest-score sites, Butte–Angels Camp, focused most of its energies on a theme of a "diagnostic and prescriptive" approach to education. Although the district had a larger than average number of components (14), approximately half of them were related to diagnosis and prescription and were not viewed as discrete and differentiated program elements. Teacher resistance to diagnosis and prescription seriously hindered its implementation and spread to the related components as well.

This association between intraproject differentiation and overall project success suggests that there are important differences between a project consisting of a few general components and one having a large number of more specific ones. We might hypothesize, for example, that if the planning process involves an assessment of organizational problems in some detail, and if components are designed to deal with specific problems that emerge, the end result will be a larger number of components. Conversely, if problems are perceived in a very global (and probably ill-defined) way, the result may be a few global and ill-defined components.

High scope scores are associated not only with large numbers of highly differentiated components, but apparently with components that meet specific, locally perceived needs. In order to determine the extent to which the formal project plans reflected widely acknowledged local problems, we compared questionnaire responses of professional staff and a sample of community residents about their perceptions of local problems with the program goals stated in the formal project plans. Four districts were given a high score on this variable of goal–problem congruence (see Table 5, p. 120). Of these four districts, Prairie Mills, Oyster Cove, and Timber River received the highest overall scope-of-implementation scores. These three districts also had the largest number of components (the fourth, Clayville, had a moderate number of components). The importance of this finding, even though it is still somewhat speculative, should not be underestimated. It suggests, for example, that breaking a complex, district-wide change project into an array of specific components which attack limited problems is more likely to result in successful implementation than is defining problems and solutions in a very global way.

This finding suggests that we may view the ideal planning process as one which breaks down the perceived needs of the system into discrete problems and produces solutions in the form of a highly differentiated, multicomponent project. There are several advantages to such an approach. The incremental effect of a number of small successes appears to give momentum to a district-wide change process. Discrete components which serve as solutions aimed at specific problems can also facilitate a formative evaluation process, since it is easier to assess their impact. Such components are also more easily modified or abandoned, especially where there are a sufficient number so that the decision to abandon any given component does not leave a gaping hole in the project. Finally, a large number of integrated components can also provide something for various political constituencies within a school system.

In addition to a large number of differentiated components, three of the high-scoring school districts have another common program characteristic. Salmon Point, Oyster Cove, and Timber River each built upon plans for change that were already under way before they received funding from the ES program. Building an ES project on previous change efforts was particularly important in Salmon Point; there, the component that received the largest proportion of the ES funds was an already existing curriculum emphasizing basic skills which was simply incorporated into the ES project. In Oyster Cove, there was a relatively active in-service training program prior to ES, which was assimilated into the ES project. A substantial proportion of Timber River's funds went to their "Instructional Leaders program" to assist principals in curriculum development—a program that had begun in a limited way earlier. Timber River was also able to continue the existing community coordinating council, which served to increase community participation in the schools.

Although the inclusion of such previously instituted activities may appear counter to the concept of ES as a program of innovation, these three districts were in compliance with the Announcement, which stated that comprehensive change "does not mean that everything currently being done . . . must be replaced completely." Rather, components in each facet must be "related to, consistent with and supportive of all of the other areas" (the Announcement, p. 2). Thus, including an activity already in place which fit into a comprehensive design was consistent with the program's guidelines and, in retrospect, highly advantageous.

The incorporation by a district of previously planned programs into its ES project has two implications. First, it indicates that the district had already had some measure of experience with the design and implementation of district-wide innovations. Even though the programs that preceded ES were much smaller in scope, a history of even modest past innovations augurs well for future large-scale innovation. Second, the existence of an already functioning component which met some of the objectives for comprehensiveness required within ES meant that these districts could devote more of their energies to improving other facets of the educational system. Thus, they had a head start.

Another characteristic of the projects affecting the total scope scores was the number of components which a district chose to implement in one year. Of the four districts which scored the highest, only Salmon Point attempted to implement all its components at once—however, Salmon Point's rate of discontinuance was the highest of the 10 districts (see Appendix A). The other three high-scoring districts implemented components sequentially over a two- or three-year period. Especially important seems to be the fact that all four high-scoring districts had staff training programs that continued throughout the three-year period of project implementation.

An additional difference between the high- and low-implementing districts was in the amount of federal money the districts received for planning. In 1972, when ES/Washington was selecting school districts for participation in the ES program, sites were chosen in two waves. First, six school districts were selected and given a one-year planning grant (ranging from $90,000 to $120,400) with a moral commitment for funding for an additional four-year period. These monies also included provisions for early attempts at implementation. Four additional school districts which were considered "less ready" by ES/Washington (and therefore higher risks) were given much smaller one-year planning grants of $46,500—with subsequent funding conditional on approval of their formal plans for an ES project.

It is noteworthy that all of the districts in which there was greater change had received larger planning grants. These districts were able to produce project plans, albeit after a number of painful iterations, which seemed to operate more smoothly as the basis of implementation. (The Spearman's Rho correlation between the size of the planning grant and level of implementation is .86, significant at the .05 level.)

One speculation that it is possible to make concerning this finding is that the screening process at the time of the submission of the letters of interest was successful in judging the readiness of the applicant districts and that the first group of six districts (each of which received larger planning grants) was "readier" for the ES program than was the conditional group. However, we may speculate that the higher level of funding for planning facilitated implementation over and above the readiness quotient of the districts. On the other hand, the level of district funding for *implementation* did not necessarily affect a district's scope of implementation (see Table 32).[6]

Contextual Factors Associated with Implementation

Although certain factors associated with the design and management of the ES projects appear highly related to their level of implementation, it must be emphasized that the ES projects were never independent of their district contexts. They were dependent not only upon support from their community and from the district professionals, but also upon the ability of these professionals to implement the various components.

An important feature of the district projects which achieved a high scope-of-district-implementation score is to be found in the level of support of local professional staff and parents for the local ES projects. Based on the assumption that staff who view the project favorably are better prepared to implement the projects, we asked the professional staff members for their perceptions of the percentage of teachers and parents who thought ES was a good idea (professional personnel

[6]The Spearman's Rho between level of implementation and funding and the scope-of-change score is .43, which falls short of the significance at the .10 level. In computing this correlation we took the level of district implementation funding and divided it by the log of the number of students in the district. The logic of this approach was that the costs of implementation might increase in a large district, but that because certain implementation costs may be relatively fixed (such as costs of hiring consultants, running some small programs, etc.), they would not increase directly with size. The correlation between the untransformed funding for implementation and the scope score is lower—.16.

Table 32. Rank Order of Level of Implementation and Amount of ES Funding for Planning and Implementation for Each School District

School district	Rank order of scope of implementation	Planning grant June 26, 1972 to June 25, 1973	Three-year contract June 25, 1973 to June 30, 1977
Timber River	1.0	$120,400	$992,210
Prairie Mills	2.0	99,250	715,611
Oyster Cove	3.0	109,000	379,773
Salmon Point	4.0	90,500	265,365
Clayville	5.0	118,000	567,505
Magnolia	6.0	91,500	518,133
Desert Basin	7.0	46,500	699,680
Butte–Angels Camp	8.0	46,500	733,840
Liberty Notch	9.5	46,500	336,953
Big Sky	9.5	46,500	361,351

census form, fall 1973). In three out of the four highest implementing districts, at least 75% of the teachers indicated that ES was a good idea, as opposed to no more than 49% in the four lowest-implementing districts. Even taking into consideration the other high-implementing district, Timber River, which reported relatively low staff support for the project (39%), the four high-implementing districts reported 66% of staff favoring the projects. Similarly, with regard to parents' perceptions of ES, of the four highest implementing districts, 55% of the parents were perceived by teachers as supporting ES as opposed to 30% in the low-implementing districts. (Once again when we include Timber River, the overall average support in the high-implementing districts was 47%.) It thus appears that staff and parent support for an innovative project increases the likelihood that the innovations will be implemented. It must be noted, however, that early support for a project does not guarantee continuation of activities after they have been implemented. Such continuation seems to require sustained and even increased support and acceptance of project activities, a situation which did not arise in many of the ES districts. In Prairie Mills, for example, many project components were ultimately discontinued, primarily because of loss of support and interest by local staff (Donnelly, 1978). (Factors affecting continuation are described more fully in Chapter 9.)

We also examined the linkages between the two major role partners in the change process, administrators and teachers, and what these linkages mean for the implementation of change. Both teachers and administrators are instumental in the process of change. These two groups, however, may be expected to function in somewhat different milieus as a result of focus of their major tasks (Cohen *et al.*, 1976). In general, administrators (superintendents, central office staff, and principals) are less concerned with day-to-day curricular or pedagogical issues and more involved with overall planning; they are likely to have more contact with the other levels of the school system and layers of the districts' sociocultural environment. For example, district-level adminstrators must negotiate with the school board and other representatives of the community over policies and funding, and, as a result, are part of the political environment and therefore susceptible to community pressure. In some instances (e.g., Magnolia), the superintendent is elected by the people at large and is even more subject to external political pressures.

On the other hand, teachers tend to be oriented to the classroom and the pupil. Their contact with the community over school-related issues is usually limited to meetings with parents and parent groups directly associated with their pupils. Teachers are very close to the daily decisions that occur within schools and classrooms but may be somewhat removed from overall policy making and planning.

These two role levels are connected by both formal (structural) and informal (normative) linkages which are likely to have implications for system coordination and for program implementation. In our examination of these linkages, we expected that roles would be less clearly bounded in these small rural districts than they are in large urban districts. Since the rural districts have only limited staff there is often considerable overlap between the teaching role and the administrative role. For example, individual teachers sometimes have responsibility for coordinating curricula across all schools within a district. In turn, principals frequently spend a substantial amount of their time teaching.

The involvement of administrators in teaching roles might be expected to lead to higher levels of shared attitudes or increased *normative* linkages. Because of their exposure to the problems and culture of teachers, administrators who have dual roles may develop more collegial relationships and, to a greater extent than single-role administrators, may depend on such collegial relationships to coordinate their school systems.

To see whether teachers and administrators in general perceive the world from somewhat different perspectives, based on their differing task orientations, we compared differences in administrator and teacher perceptions of the structure and culture of their school system by aggregating individual responses to questionnaire items given by the two different role groups to the district level, and conducting tests to determine whether there were significant differences between group means.[7] Table 33 shows that such normative differences exist. When all administrators and all teachers are included in the analysis, there appear to be systematic differences between administrators and teachers on four of the five structure variables (enforcement of rules as well as three of the authority variables) and on four of the six culture variables (problems, goal discrepancy, morale, and change orientation). However, when data are analyzed by district, we find that only Big Sky stands out as exhibiting significant differences between the two role groups on a large number of the variables. As we pointed out in Chapter 4, this district had just undergone a major administrative reorganization and consolidation, which may well account for the apparent split between teachers and administrators. In Prairie Mills, Clayville, and Oyster Cove, we find an intermediate number of normative differences; in each case three of the 11 tests reveal significant differences.

We then examined the effect of administrator time spent in the classroom on these normative linkages within the districts. There is considerable variation across districts in the degree to which administrators function in a dual role (i.e., devoting some of their time to teaching) or in a single, administrative role (see Table 34). We were able to find no clear evidence to support our expectation that districts characterized by dual-role administrators would have higher consensus between the two groups. Big Sky, with the highest level of disagreement, and Liberty Notch, with the lowest, were both districts with dual-role administrators. In Prairie Mills and Oyster Cove, which had intermediate scores on level of consensus between different role groups, administrators spent no time in the classroom; but in Clayville, which had the same intermediate score, there were administrators with dual roles.

[7]These data were available only for 1977, and only for the 11 variables displayed in Table 33. Structure and culture variables presented elsewhere in this volume, but not included in this analysis, were either not available from the administrators' version of the questionnaire, or were obtained from questionnaires administered to the OSRs.

Table 33. Significance of Differences between Teacher and Administrator Mean Scores on Structure and Culture Variables

	Structure variables					Culture variables					
T/A comparison	Enforcement of rules	School-board authority	Superintendent authority	Principal authority	Teacher authority	Problems perception	Goal discrepancy	Morale	Change orientation	Tension	Pupil autonomy
All districts	b	a		a	b	b	b	b	b		
Salmon Point[c]	—	—	—	—	—	—	—	—	—	—	—
Prairie Mills								b		b	
Big Sky	a	a		a	b	b	a	a		b	
Clayville		b		a				b			
Butte–Angels											
Camp								a	a		
Liberty Notch											
Magnolia									b		
Oyster Cove	b				a			b			
Timber River		b						b			
Desert Basin								b			
All teachers	6.08	5.88	8.14	8.60	6.06	10.43	2.55	43.65	32.71	6.39	21.34
All Administrators	8.06	5.17	8.56	9.33	7.18	8.50	1.48	34.29	24.00	5.20	21.15

[a] t-Test significant at the .10 level.
[b] t-Test significant at the .05 level.
[c] No tests done at Salmon Point due to low response rates from administrators.

Table 34. Scores of Districts on Selected Variables

District	Variable			
	Dual-role administrators	Single-role administrators	Number of significant normative differences between role groups	Implementation rank
Big sky	X		8	9.5
Clayville	X		3 ($\bar{x} = 3$)	5 ($\bar{x} = 7.5$)
Liberty Notch	X		0	9.5
Magnolia	X		1	6
Butte–Angels Camp		X	2	8
Prairie Mills		X	3	2
Oyster Cove		X	3 ($\bar{x} = 2.2$)	3 ($\bar{x} = 4.2$)
Timber River		X	2	1
Desert Basin		X	1	7
Salmon Point	N/A	N/A	N/A	4

Apparently the *administrative* role, with its responsibilities and its requirements to take a system maintenance perspective rather than a classroom focus on purely academic and curricular issues, is a major determinant of attitudes and perceptions. In addition, local factors, such as strikes (as in Prairie Mills) or major organizational changes (as in Big Sky), can have a clear impact above and beyond the degree of role segregation.

Our examiniation of the relationship between normative linkages and ES implementation revealed that there was a negative relationship between the two (Table 35). Of the five districts with high levels of agreement between teachers and administrators, four received implementation ranks that were below the average for all the Rural ES districts (Liberty Notch, Magnolia, Desert Basin, and Butte–Angels

Table 35. Districts Categorized According to the Relationship between Level of Implementation and Level of Normative Linkage and between Level of Implementation and Level of Structural Linkage

	Normative linkage: High (2 or less significant areas of disagreement)	Normative linkage: Low (3 or more significant areas of disagreement)
Implementation High	Timber River	Prairie Mills Clayville Oyster Cove
Low	Butte–Angels Camp Liberty Notch Magnolia Desert Basin	Big Sky

	Structural linkage: High (Single-role administrators)	Structural linkage: Low (Dual-role administrators)
Implementation High	Prairie Mills Timber River Oyster Cove	Clayville
Low	Magnolia	Big Sky Butte–Angels Camp Liberty Notch Magnolia

Note: Information for Salmon Point was not available.

Camp). Conversely, of the four districts that had higher levels of disagreement, three had above-average implementation scores (Clayville, Prairie Mills, and Oyster Cove). The exception is Big Sky, the district that exhibited the highest levels of disagreement, which was low on implementation. However, it is difficult to determine whether this association is related to the ES change process or to vestiges of the bitter conflicts over consolidation.

Although the holding of dual roles on the part of administrators did not seem to affect the existing level of normative linkage (whether high or low) in the districts, a relationship did emerge between single- vs. dual-role administrators and successful ES implementation: districts with single-role administrators tend to have higher implementation scores than districts with dual-role administrators. Of the three districts that had only single-role administrators (0% of administrative time spent in classroom teaching), three were above average in implementation. Conversely, of those five districts where administrators typically spent some time in district classroom contact, four were below average in implementation (see Table 35). The one high implementer among this group had a relatively small average percentage of administrator time in the classroom.

Since we are unable to explain this finding through our analysis of normative linkages, it appears that the differences in successful implementation between districts with dual- and single-role administrators lie in the realm of structural linkages. We have found (both in these data and from the case studies) that strong leadership is a crucial element in carrying out a change program. We may infer that such leadership becomes more likely when an administrator is free to devote his or her time and attention primarily to issues of system maintenance and coordination.

CONCLUSIONS

The data from this analysis of normative differences between role partners and structural linkages within districts tend to confirm and augment our earlier conclusions. We see that administrators who also have teaching responsibilities are not necessarily more likely to have attitudes or views of the school system that are similar to those of teachers in the district than are administrators who are not involved in direct classroom responsibilities. In addition, we have seen that in these small rural districts, consensus between teachers and administrators is negatively rather than positively associated with the degree ES implementation. This fact implies that normative linkage between levels within the rural school district will not necessarily facilitate implementation, and may actually impede it. On the other hand, we find a relatively strong negative association between administrator involvement in the classroom (the dual-role administrator) and implementation. One interpretation of this finding is that lack of differentiation between teachers and administrators, which is implied by a dual-role administrator, may severely decrease the administrator's influence in the rural district. This diffusion of formal leadership appears to impede the potential for change. It appears that, at least in rural school districts, structural linkages are required for successful implementation of district-wide changes, for the data suggest that single-role administrators are able to devote their undivided attention to leading successful efforts for change.

This chapter has examined the nature of implementation in school districts and has identified some of the programmatic and organizational factors that aid or impede successful implementation. Chapter 9 continues our analysis of the change process at the district level, addressing the important issue of continuation of change.

9

The Continuation of Change

An examination of the degree to which ES sponsored activities continued in the 10 rural school districts is particularly salient to this study of organizational change because of the explicit premises of the ES program. The 10 districts were to use ES funds to plan and implement comprehensive changes which would be "continuable" once the special funding ended. Thus, as described in Chapter 4, ES funds were considered to be a temporary "bubble in the budget"—special funding for a massive influx of resources to achieve the objective of comprehensive change.

This expectation presupposed that ES monies would serve as a "catalyst" for change; that they would be used to *facilitate* change rather than to sustain programs. Some changes might cost nothing at all to maintain, such as a reorganized schedule or a community advisory committee; it was expected (or hoped) that other, more costly changes would become so important to the districts that they would find local funding for them.

The ES/Washington planners did not intend to force the expansion of local school-district budgets at the end of ES funding to accommodate expensive alterations. Instead, at that time, each district was expected to be doing markedly different educational things, but at either the same per-pupil costs as before its ES project, or at an increased level which could be borne by the school district itself.

However, a recent study by Berman and McLaughlin (1977) casts considerable doubt on the assumption that schools will usually be able to continue programs that were implemented with externally provided "seed money." Their investigation of federally funded Title IV programs for educational innovation revealed that the implementation and support strategies used by many districts in launching innovative programs reduced the likelihood that such innovations could be continued unless alternative, *external* support funds could be found; thus, even successful implementation does not ensure later institutionalization.

OUR APPROACH TO INVESTIGATING CONTINUATION

Despite the fact that implementation occurred primarily at the school level, our investigation treated continuation as a district-level phenomenon, since ES was designed as a district-level program. Without exception, the early documents prepared by ES/Washington planners referred to changing "education" in all facets for all students in all grades, and conspicuously did not refer to change in individual "schools." Although schools became the unit of implementation, they were not conceived of by the ES program planners as the unit of change. Furthermore, as discussed in Chapter 4, planning of the projects and the negotiation process between the federal agency and the local school districts were at the district level, as was the responsibility for building into the projects a plan for continuation. Finally, if continuation were to be district wide, one can presume the necessity for continued administrative support at the district level, for the decision to be made and enforced at the district level, and, if necessary, for provisions to be made for funding allocation—also a district-level decision.

We also paid attention, however, to continuation of change within specific schools. A decision to continue, especially when made at a centralized administrative level, does not necessarily guarantee that the decision will be carried out at other levels in the system, such as in all schools or relevant classrooms—particularly in such loosely linked systems as schools. However, continuation occurring at anything but the district level would constitute only "pockets of continuation" and would not fulfill the ES program's objectives. Only if continuation were visible district wide would the districts be achieving the explicit objectives of the ES program.

As we have indicated in Chapter 2, the literature on organizational change has paid only limited attention to the continuation of programs after their apparent implementation. However, a number of recent works have influenced this phase of our investigation (Berman & McLaughlin, 1977; Hage & Aiken, 1970; Yin & Quick, 1977). For example, Hage and Aiken (1970) have labeled the final stage in the process of change "routinization." This very meaningful term implies that continuation is not simply a passive or automatic outcome, but that there are processes or activities that must occur to "routinize" that outcome. When one observes these processes or activities, one can be sure that innovations have become "routine" and established parts of an organization's procedures.

Yin and Quick (1977) have attempted to elaborate and operationalize the concept of routinization in a study of technological innovations. As noted in Chapter 2, they identified a number of organizational resources or operations that are necessary to continue an innovation. These include budgetary resources, personnel, training, governance or regulations, and the purchasing and maintenance of supplies. Furthermore, they posit that these resources must undergo certain "passages" (e.g., budgetary resources must be converted from soft to hard money, functions must become part of job descriptions, skills must become part of professional standards, use of innovation must become formalized in writing, and equipment must be purchased and maintained), and survive through various temporal "cycles" (e.g., annual budget cycles, turnover of key personnel, repeated cycles of training and purchasing). These factors, they maintain, are indicators of routinization. Such an approach, although not targeted to typical educational innovations, can be helpful in understanding the meaning of routinization of educational programs.[1]

Our approach to investigating continuation owes much to this literature, but we felt that certain additional dimensions were also important, both from the perspective of assessing continuation of educational innovations in general, and for establishing criteria by which to assess achievement of the ES goal of comprehensive change.

[1]Unfortunately, our investigation of continuation could not await the time delay required to determine whether some of the passages and any of the cycles took place, for it was conducted immediately after funding ended in seven of the 10 school districts, and during the final year of funding in three of the districts.

Our assessments were based, in particular, on the following dimensions (see also Table 36):

1. *The kinds of activities* that were continuing: Were they core and central to the educational process, or were they "add-on" enrichment programs? The continuation of at least some core and central activities needed to be considered as a factor in successful continuation. While enrichment programs may have had cultural value to the district, it was the intention of the ES program to transform the educational processes within the district.
2. *The fidelity of continuation:* Was the activity continuing as implemented or had it been modified? Were there spin-off ac-

Table 36. Definitions and Illustrations for Dimensions Used in Assessing Continuation

Dimensions	Definition	Illustrations[a]
Kinds of activities continued		
Core and central	Vital to functioning of the educational process. Impact on schooling considered important.	Semester system in jr. high and elective system in jr. & sr. high Independent studies (Oyster Cove) Sequenced curriculum (Liberty Notch)
Add-on enrichment	New offerings which have cultural or other value but are not integral to the educational process.	Resident outdoor school (Timber River) Adult ed. & recreation (Desert Basin) Marine education (Oyster Cove)
Fidelity of continuation		
Total	Activity continuing as implemented—in all schools where it was implemented.	Reading as middle-school subject (Prairie Mills) Travel program (Salmon Point)
Partial	Activities modified in form or continued in reduced number of schools, classrooms.	Career-Ed. component absorbed into social-studies curriculum (Liberty Notch) Teacher aides who served grades 1–6 under ES reduced to grades 1–3 (Prairie Mills)

Table 36 (continued)

Dimensions	Definition	Illustrations[a]
Spin-off	Activities which are attributable to ES or which have evolved from ES but do not represent implemented components.	ES in-service experience led to putting one in-service day on school calendar where there were none pre-ES (Oyster Cove) Steering Committee created to monitor ES project continues functioning as monitor for state-mandated program (Clayville) Changed approach to teaching attributed to individualization component (Butte–Angels Camp)
Type of decision		
Active	Indication that some commitment to continuation exists by such means as securing funds, changing staff roles to accommodate the activity, or formalization such as written guidelines.	Local funding acquired to support half-time assistant-principal position created under ES (Timber River) CETA monies acquired to support media center (Prairie Mills) Enforced use of minimum-competency standards (Timber River)
Passive	Activities appear to continue because no decision has been made to discontinue. No evidence of commitment to continuation. This type of decision is particularly applicable to materials.	Materials acquired for diagnostic center distributed to schools (Clayville) Audiovisual catalog available as resource (Butte–Angels Camp)

[a]Complete site-by-site descriptions of continued activities are found in Appendix A.

tivities that evolved from ES? Viewed against the ES goal of educational transformation, the continuation of some components in their totality can be seen as reflecting something "new" which arose from the ES project. We also took into account the fact that many innovations became modified in the process of being implemented, and that such modification tended to make the continuing activities more like what had existed before ES. Spin-off activities appeared to be still less important in terms of the educational goals of the program.

3. *The type of decision* to continue: Was the decision an active and conscious commitment to the new activity, or was it simply the

lack of a decision to discontinue? Active decisions represent a kind of commitment which can be considered more indicative of successful continuation than continuation which occurs passively (that is, without a decisive commitment to continue). Routinization factors were especially important in this regard, for their presence would represent evidence that some active decisions had been made and would serve as some assurance of continued survival. We looked for indications that the districts had secured alternative funds to pay for activities; that they had changed staff roles to accommodate the activities; and that the activities had been formalized through written guidelines or manuals.

These three dimensions were viewed interactively. Cases where local funds were decisively committed to the continuation of a formalized educational activity were deemed a greater success than a passive continuation of educational materials acquired through ES, or the active continuation of an activity that was popular but of low educational impact.

Measuring Continuation: The Districts' Scores

Data regarding the continuation of ES funded activities were collected directly from district administrators early in the 1977–1978 school year.[2] First, superintendents and principals received a brief questionnaire asking for information about the status of the components that had been part of the ES project in their district or school. (The questionnaire is reproduced in Appendix B.) Next, informal follow-up telephone interviews were conducted to determine the specific nature of the continued activities and of the decisions made concerning continuation. The criteria presented in the previous section served as the basis for these interviews.

After the interviews were completed, we selected a panel of

[2]At this time, three of the school districts were in their final year of ES funding: Big Sky, Liberty Notch, and Butte–Angels Camp. The remainder were in the first year after the termination of ES funding. Those districts still being funded were asked about planned continuation. Although implementation scores were based on data derived from On-Site Researcher Questionnaires, it was not possible to apply the same metholodology to collect data on continuation because of the early termination of the field effort. OSRs reviewed and critiqued this chapter as well as Appendix A. However, the reader is cautioned that the continuation data may reflect a bias inherent in self-reports of local administrators.

judges to rate the districts on their level of continuation. Panelists included both members of the study team who had conducted interviews and members of the study team who knew few details about the continuation activities. The seven panelists reviewed the questionnaires for each district and were briefed by the interview staff on the outcomes of the telephone calls.

During the assessment of the survival of ES stimulated changes in the 10 districts, it became clear that in no case was survival complete: in no case was there continuation of *all* ES initiated activities or changes. However, some elements of each district's project survived. Further, even when certain activities did not continue, their impacts were being felt indirectly through various spin-off activities that were taking place. The panel members independently ranked each of the 10 districts on its level of continuation. Since the rankings among judges were not perfectly correlated, in part because of the inability of several judges to eliminate ties, it was decided to classify the districts into four groups. In this way a complete consensus among all judges was achieved.

Two districts, Timber River and Oyster Cove, were ranked "high." (It should be kept in mind that "high continuation" is a relative, not an absolute term.) Three districts were ranked "moderate"—Salmon Point, Prairie Mills, and Clayville; three districts were ranked "moderately low"—Liberty Notch, Butte–Angels Camp, and Desert Basin; and the two remaining districts, Magnolia and Big Sky, were ranked "low" on continuation. (Appendix A contains site-by-site descriptions of the continued activities.)

Some Observations on Patterns of Continuation

Several general observations can be made about the patterns of continuation that were found among the 10 districts:

1. The districts that had very high implementation scores (as reported in Chapter 8) had to discontinue many of their components.
2. The districts that had low implementation scores had higher levels of continuation than the data available from the reports of the OSRs had led us to expect.[3]

[3]It must be noted, however, that the field efforts of the OSRs ended at the close of the third implementation year, one year prior to the earliest termination of federal funding.

3. All districts appeared to exhibit some permanent effects of the ES experience—in particular, the achievement of some important local objectives.
4. Certain programmatic features and assumptions of the ES program were perceived by many of those interviewed to have negative effects upon the process of implementation and continuation.
5. Our ratings of continuation were highly related to ES district staff estimates of the impact of the program on the curriculum and structuring of their schools.

One observation of our judges was that the gap between the high and low implementers narrowed as the districts moved into the continuation stage. All the districts found it necessary or desirable to drop elements of their projects once the special funding ended. In the case of the districts with the largest number of differentiated components which had received high scope-of-implementation scores (Timber River, Oyster Cove, Prairie Mills, and Clayville), attrition was greater—although in absolute terms, more project components continued than in the other districts.

This finding lends credence to the speculation that it is easier to evaluate and abandon a discrete component than an amorphous one. Districts with many components may also find it easier to set priorities among them and to streamline the project into a manageable size for incorporation into the system.

The "Building Instructional Leaders" component in Timber River, which was integral to management of the implementation process there, serves as an example of a successful component which was dropped when managing implementation was no longer a district priority. Similarly, release time for teachers in Prairie Mills, which was designed as a mechanism to facilitate implementation, was also considered successful but was abandoned when the district had accomplished the implementation process. Furthermore, many components in the high-implementation districts were actually special enrichment programs (such as camping or travel programs), and these became early candidates for discontinuation, unless they were so highly popular that school boards or parent groups were somehow able to find other sources of funds for them.

A second related finding regarding the continuation of ES components was that the low-implementing districts retained more components than might have been predicted from their past performance within the ES program. Relatively low-implementing districts (such as Liberty Notch and Desert Basin) were left with better-coordinated reading programs, more individualized instruction, and more coordinated curricula in language arts, although these generally survived as piecemeal changes in specific schools rather than as a comprehensive district-wide change.

The unexpected strength of survival shown by ES components within individual classrooms and schools in low-implementing districts may be explained by examining some functional aspects of system linkage (see Chapter 2). The low-implementing districts were characterized by especially loose linkage and tended to lack administrative coordination and strong administrative support or enforcement of comprehensive change and continuation. However, a corollary aspect of loose linkage is the autonomy it allows innovative principals and teachers to continue innovations independently. To some degree such "localized" continuation occurred in all the districts, but it is the most revealing in districts that were least successful in implementation.

Although our judges agreed that the level of continuation in all 10 districts fell short of the level of implementation, and that the continued effects of the program hardly represent "comprehensive change," they also agreed that the ES program did leave its imprint in each of these 10 rural districts and that some changes were persisting. If the overriding goals of the ES program were not achieved, some local goals were achieved. A case in point is Clayville, whose local goals for its ES project were concerned with improvements (even cosmetic improvements) which would attract new residents through the schools. As the project in Clayville wound down in 1975–1976, one principal noted that, although he did not know if the schools were now educationally superior, "We've got people thinking we are and that's the main thing." An industrial spokesman (formerly critical of the schools) agreed that the image of the schools had changed to the satisfaction of the people in power in the county; especially notable was the accreditation of the high school (Clinton, 1979a, p. 314). In Prairie Mills and Magnolia, respondents to the telephone interviews claimed that their districts had become more sensitized to local problems and needs and to methods for solving them. Still other districts, particularly low-implementing ones, appear readier for change than they were when they entered into this process. An interviewee in Liberty Notch, for example, stated that he now viewed ES as a "lost opportunity"; if he had it to do over again, he would do it differently.

Another conclusion regarding continuation was that certain programmatic features of ES hindered the success of the projects. Respondents in several districts volunteered the information that they never recovered from the tensions and conflicts of the planning year and the process of contract negotiation with ES/Washington, during which many teachers felt disenfranchised and others felt they had lost "ownership" of the project (see Chapter 5). Administrators from some districts volunteered that the name "Experimental Schools" doomed local acceptance of the projects from the start. In Timber River, for instance, the word "experimental" evoked such comments from parents as, "I don't want my children experimented on," and "What experiments are they conducting?" The superintendent was told, "You should explain to the people that the program is the experiment, not the children" (Hennigh, 1978). Desert Basin (among others) renamed ES "the Small Schools project" in an attempt to head off such problems.

Administrators in some districts stated that the projects attempted too much and went too far, both in the number of changes that were planned and attempted, and in the radical nature of the changes or the prevailing philosophy behind them. The superintendent in Desert Basin stated that the projects attempted 75% more than it was possible to do. From his perspective, such massive upheaval was unnatural and counterproductive.[4] The disruptions that were created in attempting to implement many changes, particularly with the introduction by his predecessor of many extra people in coordinating roles in the schools, was disruptive of the typical behavior patterns and relationships of principals and teachers (A. Burns, 1978).

Even the large amount of money flowing into the districts caused headaches for some local administrators, who lacked the expertise to appropriately use and coordinate such resources. One administrator in

[4]It is relevant that this superintendent (the third during the ES years) was brought into the district partially as a result of anti-ES sentiment and saw his task as one of quieting district turbulence created by his predecessors. This district became so disenchanted with the ES project that it elected to terminate its relationship with ES/Washington one year early.

Liberty Notch commented that the program was too rich for the district, that so much money at one time was more than they could handle. In his mind, ES represented "a rich orphan left on our doorstep that nobody took in."

Finally, local-district staff sometimes were overwhelmed by the implications of the themes that were adopted as the basis of their project. Most of the projects were constructed around the themes of "individualizing," "personalizing," or "humanizing" education. In several of the districts (particularly Prairie Mills), such a theme became so exaggerated during the period of implementation that disruptive behavior and even nonlearning on the part of pupils were tolerated in the name of "humanizing" (Donnelly, 1979). As time went on, retrenchment was seen as the only course, both during the period of implementation and, even more so, after the project-implementation years were over.

Our source of data for formally assessing the level of continuation of ES sponsored changes was the administrator interviews conducted in the fall of 1977. We subsequently checked our results against data derived earlier from the professional-personnel survey administered in the 10 districts in the spring of 1977. At that time, we had asked professional staff for their estimation of the cumulative impact of the ES project upon the content of education and upon the structure and organization of the schools. (The distribution of responses on these questions is displayed in Tables 37 and 38.)

We found that respondents in districts which we later classified as high or moderate on continuation were reporting more changes in content and organization than were those in districts that we scored

Table 37. Percentage of Respondents Reporting Changes in Curriculum Content by District Rank on Continuation

	District rank on continuation			
Changes in curriculum content	High	Moderate	Moderately low	Low
No changes	3%	6%	10%	23%
Shift in emphasis of curriculum elements	29%	51%	56%	40%
Addition of curriculum elements	58%	33%	27%	21%
Substitution of curriculum elements	10%	8%	4%	3%

Kendall's Tau = .09.
p = .03.

Table 38. Percentage of Respondents Reporting Changes in School Organization by District Rank on Continuation

	District rank on continuation			
Changes in school organization	High	Moderate	Moderately low	Low
No changes	3%	7%	19%	27%
Changes at the classroom level only	30%	30%	28%	28%
Changes at the school level but few changes at classroom level	12%	9%	22%	19%
Changes at both school level and classroom level	55%	54%	30%	25%

Kendall's Tau = .07.
$p = .08$.

low. For example, whereas only 3% of the respondents in the high-continuing districts reported that they had seen "no change" in curriculum content and school organization, 23% of those in the low-continuing districts reported "no change" in content and 27% reported "no change" in school organization. Sixty-eight percent of the respondents in the high-continuing districts reported that ES has caused the addition or substitution of curriculum elements, whereas only 24% in the low-continuing districts did so.

We interpret this correspondence as a validation of the scores we assigned to districts on the basis of more detailed (but qualitative) data. It is a logical assumption that changes are more likely to persist if they are viewed locally as having an impact. And, in fact, the trends in these data are clear: districts reporting the existence of ES impact are the districts that we concluded, from extensive interview data, are maintaining those changes.

FACTORS AFFECTING THE OUTCOMES OF CHANGE

Having classified the 10 districts into four ranks on their levels of continuation, we now examine factors that may be related to the outcomes of change.[5] In this section, we concern ourselves first with

[5]Rank-order correlations are presented for selected environmental and organizational characterisics with district scope of implementation. Since ratings for continuation produced only four ranks, correlations with continuation were not computed. However, given the high association of implementation and continuation, the meanings of these correlations are ascribed to continuation as well.

variables which are related to the process of change and which are associated with a "rational" change strategy. We then turn to a discussion of system-linkage variables, and consider their relationship to implementation and continuation.

Process of Change

The factor most highly related to the level of continuation is the degree to which the projects were implemented. No matter which implementation score is used (the district score based on average school scores, the district score based on aggregated component scores, or their average), if we divide the districts into groups of high and low implementers, we find that the five high implementers were the five high continuers, and the five low implementers were the five low continuers. In part, this may be attributable to the timing of our investigation of continuation. Although the high implementers, which had projects that were characterized by a relatively large number of components, may experience further erosion of their projects, our implementation scores were computed in the third year of implementation and, to a certain degree, components which failed early had already been discontinued. Thus, our scope-of-implementation measures primarily reflect sustained implementation rather than the first flush of success.

However, our rank ordering of the districts on the earlier stages in the process (readiness and initiation) are also highly related to continuation, especially the ranks for readiness. Of the five highest continuing districts, four ranked high on readiness, and of the five lowest continuing districts, four ranked low on readiness (Table 39). The relationship of initiation to continuation is less clear-cut, however, with only three of the highest continuing districts ranking high on initiation. It appears that it is the planning stage which may have the most disjuncture with subsequent activities, implying a need for more attention to the rational processes of planning. (This matter is discussed further in Chapter 10.)

Although we reiterate that the process of change is not wholly a rational one but is moderated by other factors, our data do support the notion that, in a large, complex change effort, the steps taken at each stage in the process have an effect on the outcomes of the next. Table 40 portrays the patterns of interstage transitions that the 10 school dis-

Table 39. The Distribution of School Districts by District Ranks on Readiness and Continuation and by District Ranks on Initiation and Continuation

		District rank on continuation			
		High	Moderate	Moderately low	Low
District ranks on readiness	High	Timber River	Prairie Mills Clayville Salmon Point		Magnolia
	Low	Oyster Cove		Liberty Notch Desert Basin Butte–Angels Camp	Big Sky

		District rank on continuation			
		High	Moderate	Moderately low	Low
District ranks on initiation	High	Oyster Cove	Prairie Mills Clayville		Big Sky
	Low	Timber River	Salmon Point	Desert Basin Liberty Notch Butte–Angels Camp	Magnolia

tricts went through in terms of the rank ordering of the districts in each stage. (The reader is reminded that the "highest" and "lowest" ranks are not based upon an equivalent method for each stage. High continuation, for example, reflects a fair degree of attrition when compared to high scope of implementation.)

Although no district maintained its relative position through all stages, clear patterns exist (Table 40). Most districts remained in either the high, moderate, or low groups throughout the four stages of the change process. If they moved from one "position" to another, it was typically in gradual steps, such as in the case of Magnolia, which demonstrated great difficulty in maintaining its relative strength in readiness and initiation into the implementation stage and subsequent continuation. Liberty Notch, on the other hand, began to "get its act together" during the implementation stage. The one exception to the gradual transitions, or maintenance of a clear course of activity, was Big Sky, which had a highly effective initiation period and then faltered extensively during implementation and continuation. This slump, however, can be explained by a unique feature of Big Sky. Big Sky entered the ES program desiring a "catalyst for unity" at a time when five geographically dispersed districts had undergone state-mandated consolidation (Messerschmidt, 1979). In effect, a major purpose for involvement in the project was the initiation period—an opportunity to plan a unifying project. However, the outcome of the negotiation process between Big Sky and ES/Washington produced a plan that was radically altered from what the local planners had originally intended. (It should be added that an 86% growth in population overwhelmed the district in the ES years and diverted attention from the project.)

Since the level of implementation was so highly predictive of the level of continuation, we tentatively conclude that the various programmatic factors that were found to be associated with high implementation (see Chapter 8) also facilitate continuation. The powerful effect of these programmatic factors also lends support to a rational theory of change and to the administrative or management approach which stems from rational-change theory. As noted in Chapter 4, this approach was the one taken by ES/Washington in its management of the ES program. ES/Washington's assumption that the change process was rational seems to have been a valid one.

Table 40. Relative Positions of the 10 School Districts at Each Stage of Change

Stage			
Readiness	Initiation	Implementation	Continuation
Prairie Mills	Oyster Cove	Timber River	
Magnolia	Prairie Mills Clayville Big Sky	Prairie Mills	Timber River Oyster Cove
		Oyster Cove	
Salmon Point Timber River Clayville		Salmon Point	
	Magnolia Desert Basin	Clayville	Salmon Point Prairie Mills Clayville
Oyster Cove Desert Basin Butte–Angels Camp			
	Timber River Salmon Point	Magnolia	Desert Basin Butte–Angels Camp Liberty Notch
Big Sky	Butte–Angels Camp	Desert Basin	
		Butte–Angels Camp	
Liberty Notch	Liberty Notch	Liberty Notch Big Sky	Magnolia Big Sky

Environmental and Organizational Factors

Although the planning and administration of the various ES projects had an important effect upon the degree to which the program affected district educational structures and practices, ES outcomes were also conditioned by a variety of system characteristics that were not open to simple manipulation through sound management of the local change process. The evidence from the ES districts suggests that environmental factors, in particular, had a strong impact upon the outcomes of ES and that certain structural features, associated with system linkage, were also important (see Table 41).

The environmental factors that appear to have the most powerful associations with ES implementation, and consequently with ES continuation, are all aspects of the level of *rurality* of the district. In general, the more a district's schools were isolated from each other and from the district office, the less densely populated the district was, and the further it was from a major metropolitan area, the lower the district scored on our implementation and continuation measures. Furthermore, it appears that increasing industrialization in the six years that preceded the district's involvement in ES had a very strong positive association with successful ES implementation. The proxy measure for increasing industrialization was the percentage of change in nonfarm employment. Increasing industrialization is also associated with the level of teachers' salaries (Spearman's Rho = .73), a factor which clearly reflects the ability of the district to attract well-educated staff with the cosmopolitan background shown in Chapter 6 to be important in facilitating the change process at the school level.

The ES districts were all considered "rural" by ES/Washington, which sent the Announcement to all districts outside of metropolitan areas serving fewer than 2,500 students. However, the above findings suggest that rurality is not a simple construct. Even within a population where the variability on the urban-to-rural continuum is highly constrained, the factors of isolation and economic base which differentiate the most and least rural districts are critical to the outcome of change.

The importance of the "rural factor" may be attributable to several characteristics associated with rurality. First is the degree of cultural or social connectedness with mainstream American culture, which could clearly affect local attitudes toward and readiness for change, through

Table 41. Correlation of Environmental, Structural, and Cultural Variables with District Scope-of-Implementation Measure

Variables	Spearman's Rho
I. Environmental variables	
Community characteristics	
Population	.09
Density (population per sq. mile)	.61[a]
Distance (miles) Standard Metropolitan Statistical Area (SMSA)	–.52[a]
Size of district in miles	–.57[a]
Degree of dispersion of school district	–.60[a]
Change in per-capita income 1950–1972	–.11
% change in nonfarm employment	.70[b]
School budget and administrative factors	
Per-pupil expenditures	–.02
Total school-district budget (1972)	–.27
Average teacher's salary	.51[a]
% increase in average teacher's salary	–.17
Staffing instability (staff turnover during ES)	–.18
II. Structure variables	
Superintendent authority	.47[a]
Teacher authority	.49[a]
Principal authority	.25
School-board authority	–.07
Formalization	–.08
Community influence	–.44[a]
Complexity of the district	
Number of schools	–.38
Complexity of the central office	–.07
Size of the district	
Number of pupils	.18
Number of full-time equivalents (FTEs) in the central office	.22
III. Culture variable	
Collegiality	.26
Level of tension	.64[a]
Frequency of disputes	.09
Orientation to pupil autonomy	.58[a]
Orientation to change	.22
Morale	.42
Perception of problems	.01
Goal differentiation	.18

[a] $p \leqq .10$.
[b] $p \leqq .05$.

familiarity with the types of programs or approaches embodied in the ES plans. As several previous studies have shown, the "modernity" of the schools' environment is associated with innovative behavior (Corwin, 1973; Herriott & Hodgkins, 1973).

However, in addition to social influences that may be associated with rurality, our data suggest that rurality may impede communication and coordination (linkage) among schools within the same district. For example, the population density of the community has a rank-order correlation with superintendent authority of .64, sugesting that in areas where the population is widely dispersed, there are problems in maintaining the centralized administrative authority structure that we have found to be a critical factor in the change process. Similarly, the degree of school isolation in the district is negatively correlated with the attitude of district staff toward pupil autonomy, one of the two culture variables that are highly associated with implementation at the district level. Orientation to pupil autonomy, in turn, was an underlying concept of many of the projects that emphasized individualized instruction or the "humanization" of education.

Some findings in Table 41 are notable for their lack of significance. One might predict, for example, that staff turnover during the implementation period of a very complicated process of innovation would have a serious negative impact on implementation. According to our data, however, the effects of turnover are negligible. In fact, interview data from several districts indicate that some administrators found that turnover aided rather than impeded implementation, for it frequently allowed them to select new staff members with more positive attitudes toward the innovation effort than those of their predecessors. This factor was cited as particularly important in districts where the original staff members were negatively disposed toward ES.

Furthermore, it appears that administrative turnover also had a negligible impact on sustained implementation, despite the strong influence of superintendent authority on implementation at both the school and the district level. Even though centralized control seemed critical in promoting implementation, such control often appeared coercive, and the actual departure of the strong administrator had the consequence of defusing resistance and facilitating local acceptance or decentralized decision making (A. Burns, 1979; Firestone, 1980).

We may speculate, however, that the positive impact of turnover will not persist as attempts are made in the 10 districts to solidify the gains made through their ES projects. Those districts which emphasized staff development and training will be affected when teachers who benefited from exposure to the planning and development procedures and to the training begin to leave. Where turnover is not massive, the remaining teachers and administrators may be able to socialize new teachers to the approaches that were developed through the ES projects. In several of the districts, however, turnover rates were so great (up to 25% per year) that within a very short period of time new staff members outnumbered the old. Under such conditions, effective peer socialization of attitudes positive to innovation becomes difficult.

Another "surprising" finding in Table 41 concerns the insignificant relationships found between scope of implementation and our measures of district size and complexity, both of which had strong predictive power when examined at the school level (see Chapters 6 and 7). However, the direction of the relationships is similar—for example, size tends to be positively associated with implementation, whereas complexity measures tend to be negatively associated.

The negative relationship between the horizontal complexity of the school district and implementation is particularly notable. Horizontal complexity, as measured by the number of schools, is strongly negatively associated with school isolation (−.80) and negatively associated with superintendent authority (−.18), principal authority (−.31) and teacher authority (−.46). This finding suggests again that it is difficult to create a consistent authority structure for decisions that affect school practice where small schools exist in remote areas. A "power vacuum" seems to occur in such districts which blocks the planning and decision-making elements so important for the type of *district-wide* innovation envisioned by ES/Washington.

It is notable that the association between authority structure and implementation is quite different at the district level and at the school level. At the school level, the only actor whose authority was strongly and postively related to implementation was the superintendent (see Chapters 6 and 7). When school scores are aggregated to the district level, however, we find that higher influence of all major actor groups is associated with implementation. Of particular importance is the fact that teacher authority, which was not significant at the school level, is significantly and postively related to district-wide implementation.

One possible reason for this peculiar finding is that both implementation scores and level of teachers' authority are highly variable by school within district, and the district-level measures may cloud this within-district variability. Finally, there is a marked disparity between the weak showing of culture variables at the district level and the importance they have at the school level, as discussed in Chapters 6 and 7.

The association between problems of creating linkages in rural areas and the implementation-continuation of planned-change programs is further illuminated when we examine the consolidation status of the district. The three districts that consolidated within three years prior to their entry into the ES program all ranked in the lower half of our implementation and continuation measures, while three of the four districts that have a longer history of consolidation ranked high in implemenation and in continuation (see Table 42). If consolidation is interpreted as an attempt to achieve coordination among schools in a rural area, and if we assume that the management and coordination of a consolidated system is more problematic during the years just following consolidation, this finding lends credibility to the notion that physical and structural linkages are a significant factor in district-wide implementation.

It is also interesting to examine the implementation/continuation patterns of the three school districts that had never been consolidated. Two of these, Oyster Cove and Salmon Point, were very small districts, each having only two schools. Both districts were required by ES/Washington to include in their project plan feeder schools that were not part of their formal district structure. In both cases the external schools dropped out of the project during the early implementation phase because of the difficulties of coordination of authority. The other nonconsolidated district, Liberty Notch, which had a very difficult time in the planning and early implementation stages of its ES project, was, in fact, not really a district. Rather, as noted before, it was a "supervisory union" composed of three districts that were very loosely coordinated in administration and had little history of cooperative endeavor.

In summary, the concept of system linkage again emerges as a powerful force that moderates change. The variables that are most highly associated with positive implementation/continuation are those that indicate a relatively high degree of linkage in a district: lower levels of physical dispersion among schools, a longer history of being con-

Table 42. The Distribution of School Districts by Recency of Consolidation and District Rank on Continuation

Recency of consolidation	District rank on continuation			
	High	Moderate	Moderately low	Low
More recent			Butte–Angels Camp	Magnolia Big Sky
Less recent	Timber River	Prairie Mills Clayville	Desert Basin	
Not consolidated	Oyster Cove	Salmon Point	Liberty Notch	

solidated (if the school district is consolidated), and higher levels of influence among different actors within the school system. In addition, certain environmental characteristics associated with the degree of "rurality." of the district are critical, particularly those of relatively high industrialization, population density, and proximity to urban centers.

Perspectives on Factors Explaining Continuation: A Cautionary Tale

Although we found a number of environmental and organizational factors to be associated with the level of implementation and continuation, what is even more apparent is the *lack* of similarity between districts that experienced similar levels of continuation. The two highest scoring districts. Timber River and Oyster Cove, and the two lowest scoring districts, Magnolia and Big Sky, reacted very differently to the ES program and its perceived effects. Data from the professional-personnel questionnaire administered in the spring of 1977 show that Timber River's staff displayed a consistently more negative view of the overall effect of ES and its impact on the educational functioning of the district than did those of *any* of the other districts, whereas Oyster Cove was the most positive (see Table 43). On the other hand, Magnolia, which failed to achieve many changes and which, in fact, had to discontinue its most central and popular component, continued to view ES positively, whereas Big Sky, the other low-scoring district, displayed relative dissatisfaction with its participation in the ES program.

The two highly ranked districts also differ widely in their size,

Table 43. Percentage of Staff Reporting a Positive or Negative Overall Effect of ES in Their School District by District

	District									
	Salmon Point	Prairie Mills	Big Sky	Clayville	Butte–Angels Camp	Liberty Notch	Magnolia	Oyster Cove	Timber River	Desert Basin
Reporting a positive effect										
Teachers	100	70	73	79	72	75	94	100	60	79
Administrators	100	100	100	91	86	80	100	100	89	80
Reporting a negative effect										
Teachers	0	2	6	21	21	15	0	0	40	11
Administrators	0	0	0	9	14	0	0	0	11	0

population, level of superintendent influence, community influence in project planning, principal influence, orientation to change, level of collegiality, and level of disputes. The relative rank order of these two districts on these variables is found in Table 44.

We must learn from these factors to proceed with caution in the interpretation of correlational analyses for action purposes. Although quantitative data-analysis techniques are appropriate for generating theory and conclusions about regularities in the behavior within a large population, they will not generally be as useful in predicting the behavior of a single school district. In the 10 ES districts, the unique combinations of historical and situational characteristics, and their interrelationships, are extremely powerful in explaining how very different districts took different paths through the process of change and achieved similar outcomes.

A "story" could be told about each of the 10 districts, but, given the major differences in the systems characteristics of the two top-scoring districts, and the opposite staff and community attitudes within each pair of districts at the two ends of the continuation spectrum, we limit the following description of patterns of change to four districts:

1. *Timber River.* Timber River is a relatively large school district by rural standards, serving 2,300 pupils. It includes three communities and has five schools. It is in an area which is not undergoing economic growth, and, if anything, is in a period of moderate decline. It is characterized by a rather conservative population which is generally

Table 44. Relative Rank Order of Two High-Continuing Districts on Selected Organizational Variables

Variable	Rank[a]	
	Oyster Cove	Timber River
Size	7	7
Population density	7	4
Superintendent authority	9	3
Community influence on plan	1	10
Parent influence on plan	1	10
Principal authority	3	9
Orientation to change	1	9
Collegiality	9	2
Frequency of disputes	8	1

[a]1 = highest rank; 10 = lowest rank.

wary of federal intrusions into local affairs. It attracted a very sophisticated cadre of administrators prior to the ES years, a fairly cosmopolitan and experienced group, especially the superintendent. The teaching staff, however, reflected the community norms and values more than it did those of the administrative leaders. It was the impetus of the highly skilled administrative staff (particularly the superintendent) which drew the district into the ES program. The superintendent had a vision of what he thought the school district could be, and felt that he and his staff had the expertise and knowledge to create a superior school system. He and his staff designed a very complex, comprehensive, highly differentiated, and innovative ES project which was well received by ES/Washington, but which from the beginning met with extraordinary community and teacher resistance. Nevertheless, the administrators skillfully coordinated and managed the implementation of the project and included in the project several components which can be characterized as "noncore add-ons" and which became popular with staff, students, and parents. During the course of the project years, however, conservative community norms seemed to catch up with what was happening in the schools—and during the third year of implementation, one of the components of the project, a packaged curriculum unit called "Man, A Course of Study," became a cause célèbre and the administration was forced to drop the component and retrench in other elements of the project as well. During the course of the project many of the teachers came to favor certain components, but in general both staff and community never fully overcame their initial resistance and distrust. The ES project coordinator and later the superintendent left the district before the end of the project to undertake graduate studies at Harvard. Although some changes had become routine before they left, the process of attrition was already under way and is likely to continue.[6]

2. *Oyster Cove.* Oyster Cove is a very small district (286 pupils) and is characterized by a far less sophisticated and less complex administrative staff than that of Timber River. While the project was also somewhat less sophisticated, it represented far more of a grass-roots effort than did that of Timber River, and met with less resistance from staff and community. Several early elements of the project proved highly

[6]For a more extensive description and analysis of the ES project in Timber River, see Hennigh (1978).

popular, such as the carpeting of the school library, which was a visible improvement for the school. Further, the variety of career offerings that were implemented in the high school seemed highly relevant and responsive to the needs of this logging and fishing community, and enhanced the popularity of the project. The disparity between community and school norms was far smaller in this district than in Timber River, in part because of the size and scope of the district and in part because of the nature of the participatory process used to develop the goals and components of the project. The impact of the ES project on school–community relations was positive in several areas: There were classes that allowed students to go out into the community, adult-education classes were provided, and some local community residents were employed as classroom aides. Many highly popular, community-based, career-oriented classes are continuing in the high school. Other core programs are continuing both at the high-school and elementary levels, and there is also a continuing emphasis on in-service training for staff.[7]

3. *Magnolia.* Magnolia is a rural, Southern community that had been recently reorganized to achieve desegregation. The second most disadvantaged of the 10 districts economically, one of its primary motivations for entering the ES program was to improve the fiscal soundness of the school system. One of the main problems that the district faced from the beginning was the lack of indigenous expertise to carry out an ambitious enterprise such as the ES project. Superintendent authority was relatively low in comparison to that in the other nine districts. Although, with pressure from ES/Washington, Magnolia sought help from consultants, the previously entrenched forms of school-district behavior were never modified significantly. It was a decision (if not stated in these terms) by the superintendent in Magnolia not to violate the strong local norms of kinship and consensus in order to create an "effective" program. If it meant institutionalizing expertise from outside or changing local traditions, Magnolia was reluctant, and thus failed to effectively muster outside resources. If one views the change process from a political point of view, Magnolia can be seen as partially successful, since the ES project was absorbed into the local, established set of values and the modest innovation that did occur was easily accommodated. Even though there was a visible

[7]For a more extensive description and analysis of the ES project in Oyster Cove, see Colfer and Colfer (1977).

desire to engage in rational problem solving, the traditional local culture overwhelmed such rationality. Thus, very little systematic change took place, although many individual staff members attribute a great deal of personal and professional growth to the staff-development efforts of the project. Despite the apparent lack of implementation and continuation, however, both residents and staff in relatively high percentages reported that the ES project had a positive effect on schooling, on teacher knowledge, on teacher interest in pupils, and on pupil skills. It seems apparent that, although highly willing to change at an attitudinal level, the Magnolia school system was unprepared for the change effort at a behavioral level and thus unable to utilize external resources in a productive manner.[8]

4. Big Sky. Of all the districts, Big Sky's motivation for entering the ES program was the least related to a desire for engaging in a local process of solving educational problems. The main objective was to use the project as a catalyst for unity—an instrumentality for bringing together the diverse schools and school districts that made up the newly consolidated school district. The locally developed theme for the planning year centered on "governance," which was felt to be the uppermost need in the district. This theme was unacceptable to ES/Washington project officers, who viewed governance as a "way to run a project" rather than as the essence of a comprehensive educational-change effort. At that point, serious conflict arose between ES/Washington and Big Sky, and the local momentum for the project was lost. The eventual theme for the project was conceived by ES/Washington, and the feeling in the local school district at the end of the planning year was that local autonomy had lost out to the federal bureaucracy with its dominating Easterners, of whom the residents of Big Sky were historically suspicious. The effect of massive population growth in the district associated with the boom in Western coal and uranium mining cannot be overlooked as a deterrent to regearing the ES effort during the implementation year, when the county was being overwhelmed by new students. This rapid population growth is also a probable cause for the continuation of the few ES related components which persisted.[9]

[8]For a more extensive description and analysis of the ES project in Magnolia, see Wacaster (1979).

[9]For a more extensive description and analysis of the ES project in Big Sky, see Messerschmidt (1978, 1979).

10

Conclusions and Implications

In this volume we have investigated a process of change and factors affecting the outcomes of a planned change program in 10 rural school districts. One of our main goals has been to contribute to an emergent theory of change. We did not seek to generate a new theory of change to add to the existing corpus of alternative approaches, nor did we anticipate that a study of only 10 rural school districts would be capable of resolving all of the theoretical and practical issues surrounding the change process that have been apparent for many years. Instead, we hoped to add to the existing literature on planned change and implementation an empirical basis that would better articulate and integrate existing conceptual approaches that are often presented as incompatible alternatives. Although this has been our primary goal, it has also been our hope that the results of this research would be useful to those involved in future efforts at planned change, for it is in the *application* of the findings of a research effort such as this one that a true integration of alternative theories of change can be most productively undertaken.

In this final chapter, we first present a summary of the findings of our research and the ways in which these findings contribute to a theory of change in organizational settings. We then discuss some implications of our findings for the design and management of educational change programs in the future.

CONTRIBUTIONS TO A THEORY OF CHANGE IN ORGANIZATIONAL SETTINGS

Our approach to the design and analysis of the study of the Rural ES program, as discussed in Chapter 2, identified two basic theoretical approaches that we hoped to evaluate. The first of these, the *rational approach to change,* is based on the assumption that organizational change is a manageable process which is primarily affected by sound program planning, effective choice of strategies for the design and implementation of innovations, and the provision of appropriate support for the innovative effort. The second we called the *natural systems approach;* it assumes that basic characteristics of the organizational system, rather than management and leadership of the process, are the major determinants of the success or failure of change efforts. Within this general category we identified several specific approaches, some of which were associated with the importance of specific system elements—structure, culture, input, or environment—and some of which were associated with the notion of system linkage, or coupling.

In the remainder of this section, we discuss the ways in which the data that have been presented in Chapters 4–9 allow us to assess these different theories of change.

Change as a Rational, Manageable Process

Our investigation of the degree to which change is a rational and manageable process focused on several aspects of the innovations and their settings. First, we looked for evidence that "successful" innovation outcomes were predictable, based on the degree to which an organization had successfully completed earlier stages in the process. (It will be recalled that we identified four stages: readiness, planning, implementation, and continuation.) Second, we examined program design features as predictors of change, operating under the assumption that program design was the outcome of a rational decision-making process. Finally, we examined an important element of the management approach to innovation and the role of the leader in supporting the change process.

Our finding that "innovation" appears to be an incremental process that involves a careful building of successful outcomes at successive stages of a change process supports a generally held assumption that has not, in fact, been put to many serious empirical tests in research involving multiple organizational units. Our data suggest that indicators of district readiness for change are associated with successful implementation and organizational change. The association between the implementation scores for the districts and the levels of anticipated continuation after the withdrawal of federal funding is also apparent.

It should be emphasized that none of these relationships is inevitable. Redemption is clearly possible in a district that does not complete the planning process as effectively as might be desired; and a district whose planned-change effort gets off to a good start may, in fact, achieve less success than might have been expected. Nevertheless, the finding that success over time is not an unpredictable phenomenon is a critical one. It suggests the importance of focusing on the early stages of the innovation process, an emphasis that has recently fallen into some disrepute as greater attention has been drawn to the importance of support for implementation and continuation. (Berman & McLaughlin, 1978; Gross *et al.*, 1971). In addition, it suggests that whatever effort of support is given to the change process at any stage is likely to have subsequent impacts. In the case of the ES districts, the prior commitment of a strong administrator may be the underlying factor affecting all of the stages.

Our study has not covered all aspects of the process where management of planned interventions may make a difference. The data do suggest, however, that *program design* is an area that may be extremely important in its impact on innovation outcomes. Although obvious to many, this point is nevertheless, in our opinion, very important because it is often ignored in studies which view implementation as the key stage in the innovation process.

The diffusion literature has, in many cases, investigated the characteristics of innovations that make them more likely to spread rapidly through a large population. Among such factors are: *divisibility,* the possibility of initating pieces of the innovation on a sequential basis, and/or testing out the innovation on a small population before using it throughout the organization; *trialability,* the capacity of the innovation to be altered or discontinued with relative ease should there be dissatisfaction with it; *visibility,* the degree to which both external and internal observers can see the innovation in action; and *centrality,* the degree to which the innovation clearly addresses major and agreed-upon organizational problems (Rogers, 1962).

Our analysis of district behaviors in Chapter 8 and 9 supports the assumption that program design features can make a difference in the successful implementation and incorporation of innovations. Of particular importance was the degree to which the attempt to create "comprehensive change" in the district resulted in a plan that incorporated many small innovations that were relatively separate and well defined. This feature of the program design enhanced both its divisibility and trialability. In those districts that had many discrete components, we found that components came and went, were expanded and contracted, and were initiated sequentially rather than during a single year. In sum, the existence of many smaller components seems to have provided these districts with a level of flexibility in the implementation process that greatly enhanced their ability to maintain the thrust of ES programming while tinkering with its parts. Conversely, in those districts which put all or most of their efforts into one or two major components, problems were more difficult to deal with, and the result was usually a scaling down of the intended scope or even discontinuation. The utility of many smaller components was equally clear in the early stages of continuation, where decisions could be made about maintaining functionally separate components rather than requiring district commitment to larger endeavors that often had major staffing (and therefore continuation cost) implications.

Visibility turned out to be more important early in implementation and during continuation than we had initially anticipated, largely because the highly visible changes (such as school carpeting, or field trips for students) won over many community residents who were skeptical of "experiments" that might affect their children, or who suspected that strings might be attached to all this federal money. Highly visible, popular programs were also those that had strong probabilities of being continued—even where continuation involved local expenses.

One important assumption underlying the rational approach to planned change is that successful implementation is more likely if the innovation is actually a good match to some locally determined and identified problem. As part of the planning year, the ES districts were required to undergo a problem-identification process, often involving formal needs assessments conducted by outside consultants, but at minimum involving many planning meetings with school staff and members of the community. There is little evidence to suggest that these activities produced realistic needs assessments resulting in better ES plans—in fact, our data suggest that the degree of community and staff participation in the early phases of the planning process turned out to be negatively related to successful implementation.

We must emphasize that the finding that active, widespread participation is not related to successful implementation is derived not only from the narrative data presented in Chapter 5, which suggest that much of the formal participatory process was lost in the reshuffling of priorities to meet federal funding criteria. The finding is also based on our analysis, later in that chapter, of teacher assessments of the degree to which they influenced the planning process and its outcomes during the period of initiation. This latter measure presumably takes into account any teacher dissatisfaction with the pseudo-democratic aspects of a planning period in which power was monopolized by federal project officers and local school superintendents. It is also important to emphasize again that this finding is far from unique in the literature on planned change. Contrary to popular assumptions about the importance of participatory decision making for the acceptance of the innovation by staff and/or community, studies of schools, urban service organizations, and industry continually report a negative relationship between active staff participation in decision-making phases and successful subsequent implementation (Corwin, 1973; Sapolsky, 1967; Yin, 1977).

These findings may appear to be incongruent with the findings of others (e.g., Berman & McLaughlin, 1978; Naumann-Etienne, 1974). However, it is important to note that the relationship between participation and implementation may vary with the timing of participation. In the above-cited studies, participation occurred in decisions that were made after the project was implemented rather than in plans to adopt particular innovations. This fact suggests that ownership of a project does not necessarily depend on the involvement of users prior to the adoption phase.

This finding does not imply that it is inappropriate to try to get broad-based participation in planning programs, but that participation may have been inadequately managed in the ES districts. Nor does it imply that it is irrelevant to have an appropriate match between needs and innovations in order to achieve successful implementation. On the contrary, our findings suggest that our higher implementing districts (and consequently, those districts that showed most evidence of continuing ES programs past the federal funding period) tended to produce formal ES plans that showed a congruence between the program and problems (as identified by school-district staff in a survey conducted for this research in the fall of 1973) regardless of how that congruence was achieved. However, our data suggest that participation does not automatically lead to such congruences.

Our analysis also brings us to the conclusion that an aspect of managed change that has received relatively little attention over the past few years is, in fact, one of the most significant. We refer here to the notion of organizational leadership, and the role played by the leader(s) during every phase of the planned-change process.

Our findings (as well as those of Herriott & Gross, 1979), suggest very strongly that the leadership patterns in these rural schools and school districts were extremely important in determining the course of change. Over and over again, the dominance of the superintendent was a powerful factor in determining whether the program succeeded or failed. The superintendent's pervasive influence over decisions about the innovative programs, skills in negotiating a plan with the external funding agency, ability to organize a plan that would be acceptable both to the local population and to the federal government, and willingness to provide consistent support during the often difficult period associated with early implementation were all important to a project's success.

While few studies have chosen to examine in detail ways in which leadership or executive behavior may support the change process, many recent studies have come to the conclusion that the factor of executive support may be among the most critical. Berman and McLaughlin (1977) found, for example, that both principal and superintendent support for the innovation were critical in predicting continuation and incorporation, a finding supported by Emrick (1977). Yin and colleagues (1977) examined case studies of technological innovation in urban service bureaucracies and found that chief executive support was the most critical variable in predicting incorporation, although it was not related to the degree of actual service improvement. Even more importantly, Yin's study draws explicit attention to the fact that the support of the chief executive is important even where that individual is not the architect of the innovation.

What these scattered but consistent findings suggest is that we should perhaps begin to turn our attention to investigating the ways in which the manageable aspects of administration behavior can support the innovative effort: What does support mean, when is it most critical, and how active does the leadership role need to be at each stage of the change process in order to facilitate planned change?

The Effects of Natural Systems Characteristics on Change

Our study has provided a variety of findings that support the assumption that the change process is moderated in major ways by factors that are either very difficult for school systems to manipulate, or that are generally not, at least in the short run, the targets of planned change efforts in schools. In addition to affirming a view of change as partly a nonrational process, some of our findings may be useful in developing a more integrated approach to a theory of factors that moderate the planned-change process.

As we pointed out in Chapter 2, the systems view of organizational behavior has spawned several alternative theories of the major determinants or organizational change. These theories have coalesced around various system elements, or groups of variables, that we have called environment, structure, culture, and input (staff characteristics). In addition, there has emerged a new perspective called the "loose coupling" or linkage approach to change. Throughout our analysis we have examined the relative importance of these different approaches, paying particular attention to the ways in which a linkage perspective could illuminate our understanding of system behavior. Our analysis has been complicated: we have looked at the outcomes and processes of planned change at the levels of school and district, and we have employed a variety of operational definitions for various types of linkage. At this point, however, it is possible to step back from the mass of detail and to provide a more integrated view of how these results contribute to our understanding of organizational change.

Alternative Theories of Change

One very important finding revolves around the evaluation of alternative theories of change. Our analysis has suggested quite strongly that none of the four system elements—environment, structure, culture, or input—is singularly more powerful than the others. In fact, when we were able to compare them directly, we found that they have unique and independent contributions to make to the explanation of change.

Our analysis strongly suggests that change is not a unidimensional construct. Instead we find that the processes supporting a high level of penetration of an innovation in an organization (quantity) are quite different from those supporting the implementation of activities that involve substantial changes in structures and procedures (quality). Quantity of change, measured in terms of time and number of pupils and staff affected, was best achieved in schools with a highly professionalized, collegial staff who had at least some experience with implementation of innovations. Such schools were also characterized by tension within the system.

On the other hand, changes that were significantly different than what existed previously (quality) were most evident in schools which were relatively large, nonisolated, most often at the secondary level, with low teacher autonomy and high levels of superintendent authority. It is clear that improved theories of organizational change must reflect a more sophisticated view of distinct (and possibly orthogonal) types of outcomes of the change process.

In addition, we found that *interactions* among different types of system variables contributed substantially to our understanding of implementation. Even though the data suggest that the system elements are, in fact, largely independent, these elements also make joint contributions to change through some of the important interaction terms that we presented in Chapter 7. Some of these interaction terms were particularly useful in interpreting the relative significance of variables that are somewhat atheoretical in their meaning. For example, Kimberly (1976) has pointed out that organizational *size* is not an important theoretical construct, and that no useful theories have been developed to explain why or how size affects organizations in a powerful way. Usually it is argued that size is, in fact, simply a particularly important surrogate variable—that is, a stand-in for other factors such as resource structure, control over environment, complexity, or stability. However, our examination of interactions indicates that attempting to interpret the effects of size outside of a contingency framework may lead to the development of a very simplistic notion of how increases in staff or number of clients (e.g., pupils) may affect organizational behavior. In Chapter 7, for example, it was shown that size of the school did not appear to make a great deal of difference in implementation, *except under certain conditions of staff tension.* Other similarly atheoretical variables, such as the percentage of males on the staff of the schools, were similarly illuminated when we examined important patterns of interaction between the "maleness" of the school and other structure or culture characteristics.

In summary, our analysis provides considerable support for an integrated systems model, in which no one particular system element necessarily dominates others in terms of its impact on the behavior of the system, its outcomes or outputs. The need for integration among alternative approaches assumes a more interdisciplinary approach than is usually taken in theory development. Although the study of organizations, and particularly of schools as organizations, has long been carried out with reference to such different disciplines as political science, social psychology, management science, and sociology, there is still a strong tendency to derive a theoretical approach from a single disciplinary basis. Thus, for example, sociology tends to emphasize structural variables, social psychology and management science the culture variables, and political science the environmental variables. If our 10 ES school districts and their 45 constituent schools are in any way representative of other organizations (as we contend they are), it is clearly imperative to stop arguing in favor of a single, narrow model and begin to use several selected approaches simultaneously.

This same argument in favor of integrating alternative theoretical models can also be derived from other recent works on implementation. A recent study of technological innovations found not only that two competing theories of the organizational motivation to innovate (production efficiency and bureaucratic self-interest) shared conceptual elements, but also that both were useful in explaining the incorporation of innovations (Yin, 1977). The existence of complex operations and processes related to change is disheartening if the objective is to move quickly toward a highly parsimonious theory of innovation and change; but, as we will argue below, uncovering the complexities of the process, where applicable, may have significant payoffs in terms of policy and practical implications.

System Linkage: "Tight" versus "Loose" Coupling

Our objective throughout our analysis was not to attempt to test any particular notions about the impact of the degree of system linkage on the organizational change process; rather, we wished to examine in a more general fashion the degree to which a linkage approach could help to illuminate the findings that emerged from our analysis of the natural systems approach to change.

One important feature of our approach was the attempt to examine system linkages in a variety of possible ways, based on a series of perspectives that were presented in Chapter 2. Among the most important distinctions that were drawn was the fact that linkages can exist in a variety of different forms: One can look at linkages within a particular unit, such as a school; among schools in a defined organizational framework, such as a district; between a district and higher levels of the educational system, such as a federal educational agency; or between an organizational unit and various layers of its environment.

We also defined two different types of linkage: *structural* (or bureaucratic) linkage, which is based on formal coordinating mechanisms within or between organizations, and *normative* (or cultural) linkage, which refers to more informal relationships built on consensus and mutual understandings.

Our examination of linkage within units (schools) revealed several very important findings. First, despite the contention that schools are inherently loosely linked units, we find that there is considerable variability in the degree to which they are loosely or tightly linked. Second, we found that several types of linkage appear to be significantly and positively related to successful outcomes of a district-wide planned-change program. First, linkages that are developed through the existence of a centralized, formal authority structure are critical: the more centralized the decision-making system, and the lower the level of teachers' classroom autonomy and influence over decision-making, the greater the likelihood of successful implementation. Of particular importance is the fact that high levels of autonomy are negatively related to change. This finding suggests very strongly that *bureaucratic* (structural) linkages are supportive of the change process and its outcomes.

In addition, we find that *normative* (cultural) linkage is critical as well. The elements of normative linkage that appear important are, in particular, the existence of high levels of collegiality and support, and the presence of overt disputes in which disagreements between staff members are openly shared and resolved. Another type of normative linkage—consensus among role parterns—is not, however, significantly related to change, although it is frequently believed to be an important prerequisite of the change process. The results of our interaction analysis (Chapter 7) further suggest that the simultaneous presence of both types of linakge is critical: significant interaction terms (superintendent authority/collegiality and autonomy/collegiality) indicate that where *both* strong bureaucratic coordination and high levels of social supportiveness are present, relatively high implementation occurs. Thus, when it comes to innovation, normative and formal systems of coordination are not, in fact, alternative mechanisms to achieve the desired end, but rather mutually supportive conditions for change.

The role of bureaucratic linkages between the schools and the district's central office is also critical, as demonstrated in our data by the strong influence of the superintendent over the implementation of change programs at the school level. It should be emphasized, however, that we suspect that our findings about the relative importance of superintendent influence over principal influence are strongly affected by the rural nature of our sample of schools. Because of the small number of schools in a district, it was feasible for the superintendent to keep tabs on activities at the school level in ways which cannot easily occur in more complex and larger districts.

Thus far in our discussion the findings suggest that loose linkage does not promote change in schools and school districts. However, there are some indications in our data that the answer may not be as simple as stating that tight system linkage promotes innovation. First, we find that looser coupling among units can effectively allow (if not encourage) local adaptation in two ways. First, it must be recalled that even in the lowest implementing districts, at least one school stood out among the rest as a high implementer. Under conditions in which all schools changed (or failed to change) in systematic ways together, these localized success stories could not have endured. Furthermore, we believe, on the basis of our interviews with local administrators, that although loose linkage does not promote district-wide levels of continuation of programs, it may facilitate continuation of *some* activities in individual schools.

Second, there is some indication that, although the role of bureaucratic linkage between school and district office is critical at the beginning of a major change effort, it is perhaps less important at later stages. We have reported elsewhere, for example, that schools in which superintendent influence decreased over the five-year ES period were more likely to be high implementers than those in which superintendent control remained high or increased.[1] This fact suggests that there may be cycles of linkage between district and school that are most effective in supporting change: at periods of initiating change programs centralized control and coordination are critical, but as the implementation process gets under way, the superintendent's control must be diminished in favor of more decentralized decision-making strategies.

Linkage among various layers of the environment and the school has also been shown to be important in our study. In particular, we find that the loose linkage characteristic of isolated schools in districts of low population density produced problems in maintaining district-coordinating structures necessary for implementation. This burden of rurality is a difficult one to overcome, even in the context of a program that was designed expressly to help rural schools capitalize on the strengths of their rural environment. From the beginning of the program, when it became clear that sophisticated negotiations between ES/Washington and the district office would shape the character of the local programs, to the later findings that districts characterized by demographic patterns that we used as indicators of rurality were less likely to score high on our measures of district-wide comprehensive change, degree of rurality was an adverse factor.

IMPLICATIONS FOR THE DESIGN AND MANAGEMENT OF EDUCATIONAL CHANGE PROGRAMS

In our effort to expand upon existing theories about organizational change, we have identified a number of factors that have implications for the design and management of change programs. First and foremost is the overall conclusion that both the rational and the natural systems approaches to change help explain the change process. We have found that change is a manageable process which can be affected by sound planning, effective program design, and appropriate support for the innovations. However, we also found that basic system

[1]This finding is based on a secondary analysis that was performed on the data gathered by this study. Although not reported in this volume, the findings from that analysis are available in Louis, Molitor, and Rosenblum (1979).

elements—the organization's environment, structure, culture, and input—as well as the nature of the linkages within the organization and with the layers and levels of its environment also determine the outcomes of the change process. The natural consequence of these findings is that the effective design and management of change require careful attention to factors most amenable to management, but also recognition of the more difficult-to-manipulate characteristics, which may be barriers to a change process if they are not identified and taken into account.

We conclude this volume with an array of implications which may be drawn from these findings.[2] Since ES as a unique program has ended, we cannot make recommendations that will guide its future. However, there is clear evidence that individuals at both federal and local levels will continue to initiate new efforts for school improvement, and in this process they will encounter a series of decision points. Although rarely in any one program are choices faced and resolved in a manner that provides clear directions to those who plan and implement other programs, there are often persuasive parallels from one program to another. Such is the case with the ES program, and it is our goal to use our findings to draw implications for individuals at both the federal and local levels who are interested in the future design and management of educational change programs.

A note of caution is in order with regard to the generalizability of our findings and the implications which may be drawn from them. These findings are most applicable to *large-scale innovative efforts,* and particularly those which are planned and implemented in a partnership between a local and an external agency. Even though we feel that many of our findings are applicable to urban and suburban as well as rural schools, we must also point out that the rurality of the school districts which we studied was a factor in many of the outcomes and may therefore limit generalizability. Even so, one characteristic of rural

[2]The reader should note that we are not presenting implications for policy or practice in the most commonly used sense of these terms. This study was not designed to provide education practitioners with information about the usefulness of particular innovations or about the implementation of particular innovations in their classrooms. Nor was it designed to compare large-scale, federally funded strategies for change (as did Berman & McLaughlin, 1978), or alternative strategies used within a legislatively mandated program in order to make recommendations for revised guidelines and procedures—such as current studies of Title I allocation policies (Vanecko, Archambault, & Ames, 1978), or of alternative programs to achieve desegregation (Royster *et al.*, 1979).

school districts which we have found to be important, that of isolation, may also play an important role in urban and suburban districts. The effects of the physical isolation of rural districts may be compared to those of the social or cultural isolation that typifies many schools in more metropolitan areas.

We hope that the discussion which follows will benefit future attempts to design and manage change efforts. Our presentation is suggestive rather than exhaustive with regard to the practical applications of this research; what we have attempted to do is to select implications which we see as especially important and which the ES experience distinctively illustrates.

Implications for the Design and Management of Change at the Local Level

Knowledge of the System as a Prerequisite for Change

Managers of change must know not only what characteristics of their system need to be changed; they must also consider which characteristics will promote and which will impede a program of innovation. If many local conditions are "wrong"—that is, if they will act as barriers to change—it may be worthwhile to work on system adjustments to improve the organization's basic health and receptiveness to innovation before plunging into a full-scale process of planned program change.

Several findings from our research illustrate the practical importance of a system-oriented perspective for designers and managers of change. For example, we found that high levels of staff collegiality and high levels of tension and/or disputes were conducive to successful implementation. These related findings suggest a first step in approaching a program of innovation: developing strategies for increasing collaborative activities among teachers and between teachers and administrators, and in that process installing mechanisms that will increase the degree to which staff members at all levels may openly share disagreements about system goals and operations. More frequent staff meetings, the creation of staff teams, and the installation of formal coordinating structures are only a few of the devices that may increase the interdependence of staff members and their consequent willingness to express and resolve conflicts.

Another of our findings was that the presence of relatively cosmopolitan staff members who have strong professional commitments (represented, for example, by degrees held and books read) facilitates innovation. The implication is that before initiating a complex change process, it may be useful to encourage staff development or to recruit staff members with the desired backgrounds and characteristics. However, the "replacement" approach to change, as opposed to a staff development approach, must be viewed with extreme caution for a number of reasons. First, given recent developments affecting the educational profession (including increasing retention of females in the work force, tax caps, and declining enrollments), the feasibility of using a replacement approach to facilitate change is questionable in many cases. Even where feasible, however, it is important to avoid another problem that may impede change—the development of two cultures among the teachers as a consequence of radical changes in recruitment policies (Corwin, 1973). Local managers of change processes should not, however, throw up their hands if their staff is far from the cosmopolitan and professionalized group that may best carry out the change process. Early evidence from a current study of locally based school-improvement processes indicates that a well-organized set of change activities can, in fact, be one of the most important sources of staff development in creating a more cosmopolitan view of education (Louis, Rosenblum, & Molitor, 1981).

The Importance of the Planning Process

Many of our findings point up the importance of the planning process as a predictor of the outcomes of a major innovative effort. The relationship between effective planning and subsequent programmatic success lends support to the notion that change is a manageable and rational process. Indeed, careful planning should be able to anticipate the effect of nonrational elements in the system that is to undergo change. The evidence is compelling that planning a program which responds to the perceived needs of the district makes the implementation process easier.

Two deterrents to effective planning in the context of the schools and districts involved in this program should be noted. First, few of the school districts studied here had experience with systematic planning—particularly planning that covered multiyear periods. Simply adding additional funds and time for planning did not provide the districts and schools with the resources that they needed in order to use their time most effectively. Although the ES/Washington project monitor, in some cases, attempted to provide some technical assistance with regard to planning, in most cases this was an insufficient resource for the learning of new skills and procedures.

These districts, while poorer in resources than the typical urban or suburban system, are different only in degree rather than in kind. Experience with long-range planning for program innovation (particularly broad program implementation of the type envisioned in the ES program) is not a customary activity in the educational arena. One of the lessons for managers of change activities that may be drawn from this study is that, in most cases, school systems would be well advised to look for external resources to assist them in developing planning skills and strategies, rather than assuming that these can be quickly developed internally.

Although ES/Washington did well in giving the school districts an entire year for planning, the districts may have come to believe that at the end of that year, planning should stop and implementation begin. Our research indicates that this belief was mistaken; instead, planning must permeate the entire change process. After they select and adopt an innovation, planners must set up guidelines that are specific to implementation and must develop mechanisms for feedback. Even the best devised plans will have to be recast in the face of obstacles or resistance. Finally, long before the end is near, planners must set up ways of continuing innovations and absorbing them into the system. Useful measures might include gradually allocating funds and staff required to sustain the innovations, rewriting guidelines and job descriptions to reflect the program changes that have taken place, and continuing to provide training or supervision, if necessary.

Effective Program Design

Although implementing many changes simultaneously may be difficult, we found no evidence to suggest that a school district can deal with only one change at a time. In fact, an attack on many fronts at the same time appeared to facilitate the process, perhaps by stimulating the system or diffusing resistance. The most effective program designs had multiple, discrete components simultaneously implemented and targeted to address the specific problems identified during planning.

Potentially, the introduction of too many overly massive changes at once may result in chaos. Indeed, reservations were expressed in practically all the districts about this possibility. But program designers need not forego the advantages of a multicomponent approach for fear of resistance to change. They can stagger the introduction of groups of components over a period of several years. They can "co-opt," or incorporate into the project, change efforts that are already established in the school district. Co-opted components impart momentum to a project and serve as a sign that the district is not to be transformed beyond recognition. Not least important, designers can build into the program some highly attractive and visible program elements which have the potential of being very popular with staff and/or parents. Although such program components may be costly and siphon off funds from the more central elements of the program design, the investment may be worthwhile in that it may help create a political climate more amenable to the entire change effort.

Details of well-designed projects will vary from one school district to another. However, in the ES school districts, certain project components seemed to display a natural affinity: curriculum change and changes in the use of time, space, and facilities, for example, or curriculum change and in-service training programs. The establishment of *continuous* in-service training was particularly valuable, since ongoing training tended to confer continuity on a district's change effort.

The Problematic Issue of Participation by Constituencies

The issue of broad-based participation in planning for change continues to present a dilemma. Our evidence suggests that such participation is extremely important but must be managed with great care and sensitivity.

It is instructive to review the experiences of several districts that encountered problems with participation. In these districts, groups (especially teachers) felt disenfranchised either because no input at all was solicited from them or because input was solicited and then ignored. The implication is that program managers must carefully delineate what they are asking of constituencies (input, endorsement, evaluation), and must take pains to use (or to explain why they do not use) what they obtained.

There would seem to be no easy way to manage the participation of constituencies. Problems with constituents were aggravated in those ES districts where the local superintendents were not aware of the extent to which they would have to negotiate their formal project plan with ES/Washington. In the process of gaining acceptance of the project plans, many of the suggestions of other contributors to the earlier plans were revised or even eliminated. Unfortunately, this kind of uncertainty about the outcomes of a participatory planning process is likely to be typical of change efforts which are externally funded and require approval of the funding agency. It is therefore extremely important that managers of change not unduly raise the expectations of constituents during the planning process: overselling the change program at first can undermine its chances of success later.

Even if this pitfall is circumvented, the nature of the linkages among teachers and administrators within the system and community members from outside the organization can represent formidable problems for management. Ways should be found to defuse resistance and build support in order to ease the way for successful change. For example, holding discussions with staff members about the potential advantages (to them and to others) of participating in the planned-change effort, particularly if they will be called upon to do additional tasks (such as paper work), may be effective. Too often, teachers and administrators are required to pour time and energy into a change effort without any clear understanding of what the effort may accomplish.

Strong Administrative Leadership

Administrative commitment to a program plus administrative time and energy invested in it appear to be the *sine qua non* of success in bringing about change. Furthermore, our evidence suggests that changes are more likely to take hold if administrators have a *distinctive* role—there must be clear role segregation between administrators and teachers. Effective leaders do not have to hold the same view of the world as teachers, or act as teachers' "buddies," or share teaching responsibilities. Instead, they must be able to provide coordination and support for teachers' activities.

This kind of leadership needs to be rationally planned for from the outset of the program. Administrators should be allocated enough time and resources so that they can turn their undivided attention to promoting the change effort. Our research also implies, however, that leaders must be sensitive to the nonrational elements with which they must deal and to the linkage role which they must play to ensure success. Those superintendents (notably in Clayville and Butte–Angels Camp) who attempted to bulldoze through the ES project aroused strong resistance and hostility among their staff.

Most importantly, the administrator must remember that the introduction of a change program can have important unintended consequences which may have impacts on both the level of implementation and the culture or climate of the system. As school people respond to a major change program and begin to alter embedded patterns and practices, resistance will arise and turmoil will be almost inevitable. This is, at least to some degree, true even in systems where individual users of a new approach or program are not resistant and planning may be adequate (G. Hall & Loucks, 1976). Improperly attended to, turmoil may lead to increased tensions among different actors in the system.

The case studies of the ES program are peppered with instances of frustration, formal grievances, and silent undercutting of the change effort. This turmoil can also lead to a reduction in community support for either the schools or the program (Herriott & Gross, 1979). It is perhaps this tension, which was prevalent in the ES schools and districts, that leads to our finding that decision-making influence is a zero-sum phenomenon. If tensions between administrator and teacher groups were resolved with dispatch, we might find that an increase in the influence of any given party would not depress the association between influence and change among other parties. In sum, it appears that a strong administrator ideally combines firm management of the rational-change process with flixibility in dealing with nonrational, often political elements in the organization and its environment.

Autonomy and Professionalism

Our data indicate that there is a strong negative association between the degree of classroom autonomy exercised by teachers and the implementation of comprehensive change. One possible interpretation of this finding (particularly in light of the finding that administrative leadership is critical) is to assume that taking control over decision making away from teachers and giving it to administrators will produce more innovative schools. This conclusion, however, is inappropriate for both theoretical and practical reasons.

First, the clash between teacher autonomy and innovative efforts that are intended to cover multiple classrooms cannot be easily resolved by simply taking control away from the teacher. The use of most educational innovations requires both commitment and sound judgment by the user (see, e.g., G. Hall & Loucks, 1976). The attempt during the past 10 years to develop "teacher-proof" innovations has, in fact, been a failure. However, it is hardly reasonable to expect that a more professionalized staff will respond to increases in centralized decision making by devoting the extra energies that are critical for program success. Thus, unless it is possible to reduce teacher autonomy while maintaining a commitment to innovation, such a strategy is bound either to fail or to produce a punitive system that requires considerable energy in monitoring teacher behavior.

In addition, the ability of administrators to make substantial decisions about classroom autonomy is becoming more limited as the unionization of teachers becomes a fact of life in most districts. Increasingly, we can expect that the boundaries between administrative and teacher involvement in decision making will be drawn up through contractual means rather than emerging through unilateral decision making or even ad hoc, bilateral negotiations.

Thus, a school system that is concerned about the ways in which classroom autonomy may be impeding its ability to produce the best-quality education for its pupils must look for noncoercive methods of producing interdependence among its parts. As we observed in Chapter 7, a major impediment to implementation was a lack of experience with collaborative structures, which was often reflected in high levels of teacher autonomy in the classroom. A reduction in autonomy through increased collaboration among teachers appears to be a far more viable route than the unilateral assumption of decision-making responsibilities by administrators (see, e.g., Keys & Bartunek, 1979).

Implications for the Design and Management of Change at the Federal Level

The design of the ES program was a significant advance over previous federal efforts to stimulate change at the local level. It emphasized multiyear funding (the five-year "moral commitment"), locally developed solutions to locally perceived needs, and both formative and summative evaluation as elements of a comprehenisve research strategy. However, the program promised much more than it was eventually able to deliver. Nevertheless, several of the findings of this study point to potentially beneficial modifications of the federal role in local change, and others illuminate (although they do not answer) a range of continuing dilemmas which are inherent in designing programs for change.

Better Definitions of Goals and Objectives

If federal partnership in local change efforts is to be effective, there must be adequate program planning at the federal level: what is particularly needed is a well-thought-out articulation of program goals, objectives, and procedures. The Rural ES experience indicates that, while the program designers prescribed a time period for local planning, they may not have paid adequate attention to planning the program at the federal level.

The absence of explicit and consistently articulated federal expectations is so often cited as a root cause of program malfunction as to be a truism. Nevertheless, much of the confusion, frustration, and anxiety of the school districts, especially during the planning and early implementation periods, was caused by ES/Washington's inaccurate assumption that a clearer sense of the meaning of "comprehensive change" would emerge with time. Although originally conceived as a research program to be funded by grants, implementation of ES by the districts took place as a "series of demonstrations" (Doyle *et al.*, 1976), administered by formal contract. Moreover, the contract was developed through "successive approximations" by the districts in negotiation with ES/Washington—a yearlong process that, in several instances, resulted in drafts being returned by ES/Washington with little more than a statement of rejection a few sentences long. ES/Washington's inability to provide a clear or consistent sense of the attributes of an adequately comprehensive plan, compounded by rigid, arbitrary deadlines, forced the districts to focus on complying with poorly articulated requirements rather than on developing broad-based support for programs appropriate for their local needs and capabilities.

The process of formulating goals and objectives necessarily requires a delicate balancing of technical capabilities and political realities, perceived under conditions of uncertainty and evaluated against vastly different individual standards of acceptable risk. As a consequence, it may be neither possible nor desirable to accurately identify the likely outcomes of a particular program. This study does not, and indeed could not, provide clear-cut directives which would result in the right balance of flexibility and specificity of goals and objectives for large-scale change efforts. The study does, however, define two outcome measures which, with further development, could help clarify the expectations for a change-oriented project at various stages of its existence. "Scope of implementation" and "dimensions of continuation" are relatively clear, replicable measures of change and the persistence of its effects in educational settings. The two scales provide specific dimensions of change which could be used during program formulation as well as evaluation, by enabling decision makers more clearly to express their objectives as outcomes. Further refinement of these measures through additional research will increase their utility to decision makers.

Better Understanding of the Target Group

Rural schools were selected as the target group for the ES program in response to the realization by many at the federal level that rural communities were a "forgotten" constituency. And yet, in spite of publicly voiced, newfound sensitivity to the needs of rural school districts, the only concession made by the ES designers in response to the relative lack of experience of rural districts with competitive grant applications was the use of a brief letter of interest for the initial selection process, and the extension of the normally 1–4 month contract development period to a full year.

The designers' explicit and implicit assumptions about the needs and capabilities of rural school districts were often inaccurate. As pointed out in Chapter 5, ES/Washington may have made some unwarranted assumptions (especially in the planning year) about the nature of rural school districts. Lack of adequate grounding in the rural context of the ES projects repeatedly created problems for the federal management of the program and soured the federal-local relationship.

The 10 rural districts selected for participation differed significantly in extent of isolation, sophistication, and previous experience with federal programs, not to mention size and budget. On the whole, personnel in these small districts did not understand the planning or contract-negotiating processes well enough to make full use of the planning year. Furthermore, most lacked the sophistication to acquire independently the outside expertise that would enable them to design or carry out effective projects.

Such locally significant differences point to the need for better congruence between, on the one hand, the needs, goals, structures, and capabilities of potential funding recipients and, on the other hand, the objectives, requirements, and resources of federal programs. The availability of federal funds may not provide enough of an incentive for participation if, in order to receive the money, a school district is required to act in a manner contrary to its political or cultural makeup.

One mechanism for more accurate understanding of potential target groups would be to create a panel of in-house or outside consultants who would be familiar with the specific needs, technical capabilities, and cultural, historical, and political concerns of the target group. In the case of ES, a group might also have assumed responsibility for developing training manuals and structuring federal reporting requirements so as to make the budgetary and fiscal responsibilities of the districts more comprehensible and, therefore, more likely to be fulfilled.

The Role of the Federal Program Officer

Because the ES federal-local partnership bypassed the state or any other intermediary agency, the program officer as the direct representative of the federal agency was put in a position of great responsibility. This role, however, lacked clarity of definition (e.g., it was unclear whether the role was one of *authority* or *facilitator* of local actions) and consistency of execution, to the confusion and frustration of all 10 participating districts.

Many districts resented greatly what they perceived to be arbitrary and heavy-handed treatment by their federal project officers, whom they variously referred to as "Harvard types," "Eastern intellectuals," and "Washington bureaucrats." In some instances such behavior was the result of shifting priorities in Washington (D.C.), but it often resulted from an ignorance about rural culture on the part of the project officers themselves, or from their inability to insulate themselves from the pressures of federal bureaucratic crises to a degree sufficient to buffer local school districts from the adverse effects of such crises.

The roles and the problems of program officers in the ES program are discussed and explored by Corwin (1977), who makes several recommendations which stem from the ES experience. Among these is the importance of *clarifying the role* of the program officer, both for the program officer's benefit in carrying out the duties involved and for the sake of the local districts in understanding what they can expect.

A major source of the confusion in the program officer role arose from the lack of consistent guidance as to how the program officer was to balance his monitoring, advisory, and technical-assistance roles (see Corwin, 1977). A possible formulation which suggests itself from this study is to establish the program officer as responsible for monitoring and providing direct assistance in the accomplishment of federally required activities (such as General Accounting Office accounting requirements), while providing to the district access to various individuals and organizations who could provide technical assistance in more substantive areas, such as strategic planning and specific educational innovations. In light of the finding that most districts were unable to locate adequate outside technical assistance by themselves, such a "linking" function for the program officer appears valuable and merits future study.

In general, the program designers seem to have overestimated their ability to maintain control over implementation of the program. This problem was aggravated by the lack of continuity in program officers during the project years. Three of the ES districts were assigned four different program officers in the course of the project. One district had six. There were interim periods when there was no program officer at all. Especially since the role was not clearly defined, turnover caused changes in style and direction of monitoring according to the personality, philosophy, and professional background of each incumbent. One remedy, suggested by Corwin (1977), was that the high turnover in programs such as ES be countered by financial incentives to those who "stick it out" until the completion of the project.

Flexibility of Time and Money

The districts involved in the Rural ES program were given an unusually long "moral" commitment of time and funding. This was intended to free them from annual cycles of refunding and to allow them ample time to plan (the original one-year planning grant) and to implement their local designs.

In the Rural ES program, at least, the long duration of the projects had several positive impacts. Beyond the obvious benefits of the planning year and the time for sustained implementation, there were unanticipated benefits as well. Where resistance to the project was encountered early in the implementation process (often as a residue of the planning year), there was time for it to be defused and to convince staff and community of the value of continuing to implement the program. Even when there was hard-line resistance, the process of staff turnover sometimes filtered out dissidents and brought in staff more favorable to the program's objectives.

However, the need for greater flexibility of deadlines for completion of plans and reporting documents, as well as for the expenditure of funds, was also made apparent by the findings of this study. The design of the ES program was based upon the assumption that school districts lacked adequate resources in order to accomplish innovation and that a "bubble" of funds, if introduced into the system, could be productively absorbed and transformed into comprehensive change without significantly disrupting the fundamental activities of the district. What transpired in many instances was that the rapidly introduced "bubble" threatened to become an "aneurysm," and the average school district, rather than give itself over entirely to comprehensive change, engaged to a large extent in less ambitious activities.

Whether the average school district could have accommodated more change of a more radical nature had the funding been structured more flexibly than it was remains a topic for further study. In particular, future multiyear efforts by federal agencies to stimulate change could attempt to use "carry-overs" more extensively in order to facilitate more natural change, as opposed to change as an artifact of the period of funding. In light of the studies by Berman and McLaughlin (1978) and the Ford Foundation (1972) showing the long-term ineffectiveness of short-term "seed-money" efforts, it seems particularly useful to suggest that federal agencies attempt to develop discretionary funds which would not be subject to fiscal-year accounting. Such a fund could be managed in the same way as the Department of Housing and Urban Development Community Development Block Grant Fund, and dedicated strictly to studies of long-term change.

The Appropriateness of the School District or the School as the Unit of Change

There is much controversy regarding the most effective leverage point for federal efforts to assist with local school improvement. In the past, many curricular reform efforts tended to focus on the classroom teacher, only to find individual teachers powerless to sustain districts. Other efforts (particularly ESEA Title III) focused more on individual schools, only to see such initiatives fail to spread beyond those schools which were being funded.

Although this issue remains unresolved, in districts as small as those involved in the ES program, the district was certainly the most practical unit of change in fiscal and administrative terms. The important role played in implementation by strong leadership (usually embodied in the superintendent of a rural district), as well as the importance of administrative support for continuation, supports the district-level approach. Offsetting these factors, however, is the problem of implementing within individual schools innovations designed at the district level.

Certain characteristics of rural areas emerged as important limitations to the district-wide projects. The physical isolation of the schools from one another and from the central office was a deterrent to successful implementation; the institution of strong linkage mechanisms to lessen isolation during a program of change might be a counterdeterrent. Furthermore, many rural school districts exist in their present structure as a result of a formal consolidation process; the resulting entities often cross natural community boundaries. Thus, consolidation has thrown together communities or former school districts that used to be "rivals" in sports or in other areas of education or community life. In such situations, an attempt to treat the entire district as a unit of change may be foolhardy and counterproductive.

What emerges from the study is that both the district and the school can and should be included in programs designed to elicit change. The district's control over resource allocation and its ability to coordinate simultaneously the efforts of many schools determinably affect the feasibility of comprehensive change, whereas the schools' ability to mobilize teachers in support of particular activities determines the likelihood of successful implementation.

The Selection Process: Funding the "Neediest" or the "Most Promising"

Congress seldom appropriates funding for its programs at a level sufficient to fund all eligible applicants. Thus, program managers often face a dilemma: Should the funds be allocated to those most in need of assistance or to those most likely to use them productively? Such a decision depends upon the objectives of the program, the resources available to the agency, and the social and political consequences of program failure for both the recipients and the agency.

This study and others (e.g., Berman & McLaughlin, 1978) have identified several factors which appear to influence the ability of schools or school districts to respond to federal efforts to induce change. The selection of appropriate recipients—both as a class and as members of a particular class—can be improved through careful evaluation along the following dimensions: the congruence of the needs, goals, and capabilities of the potential recipients with the objectives, requirements, and resources of the federal program; and the possible consequences for the potential recipients of successful or unsuccessful implementation of the program from the sponsoring agency's point of view.

This is not meant to imply that future efforts to stimulate change should only be applied to schools or districts which appear likely to be successful. On the contrary, the federal government's priorities in supporting educational improvement should continue to include equity goals and the promotion of equality of educational opportunity (i.e., the support of programs in "needy" school districts). However, our findings suggest that recipients exhibiting certain characteristics may require different kinds of assistance in order to successfully implement particular changes. Although aware of different districts' capabilities, ES/Washington was by no means consistent in its response to interdistrict differences. Although a special effort was made to select both "low-risk" and "high-risk" districts, the program made no apparent systematic attempt to assess the different amounts of technical and financial assistance needed by districts at different states of readiness. Since few districts are as ready as the most able to adopt and implement innovations, it would be valuable for future demonstrations to include recipients at both ends of the readiness spectrum, although program sponsors should not expect to treat all recipients in a similar manner. Data gathered from such studies would be useful for local district officials attempting to independently implement change.

The Role of the Federal Government in Supporting Change: "Top Down" or "Bottom Up"

Current discussion regarding the appropriate federal-local relationship in federally supported change efforts can be characterized as a debate between proponents of a "top-down" or a "bottom-up" approach. One view, largely based upon the conclusions that have been extrapolated from the *Study of Federal Programs Supporting Educational Change* (Berman & McLaughlin, 1978), is that "bottom-up" or homegrown remedies for educational problems are best. This view is bolstered by observations that many school districts are "opportunistic" in their use of federal categorical grants, and that projects initiated under these circumstances rarely become incorporated into ongoing district functioning after the termination of the grant. The frequent interpretation of this finding is that changes based solely on local agendas are likely to be more enduring and have more impact.

The alternative, "top-down" view is not, however, without its continued support. Local schools, some argue, do not have the capacity to make major changes without external direction because they behave as partially closed systems. Such a view is based on the considerable evidence suggesting that many locally initiated innovations are never implemented, or if they are implemented, they involve such minor changes in practice or structure that they have little hope of affecting students. Many who hold this opinion are also of the belief that one of the most effective ways to stimulate local innovation that meaningfully grapples with major needs is to both impose constraints (through categorical funding or demonstration grant funding) and to provide schools and educators with an array of tested, proven new programs that deal with current problems.

Our study sheds light on this continuing dilemma in several ways. While the ES program designers did not take a direct stand on the above-stated issue, they assumed that, given the abundance of educational innovations already in existence, the districts could easily choose among them and devote most of their attention to the change *process per se.* Built into and funded by the program were a planning stage, a period for implementation which the designers believed was long enough to let the projects jell, and a requirement that the districts plan for assuming the responsibility of continuation.

However, the program designers overestimated both the districts' awareness of the range of innovations available and their ability to seek out and make use of consultant help. Thus, although the focus on change as an outcome may be viable, the process of producing change may require more technical assistance than was available in the 10 districts.

Such findings as the importance of the perception of local "ownership" of the work plan and local inability to locate adequate technical and substantive guidance point to the need for a particular cooperative relationship between the federal and local levels in which resources and guidance flow "downward," while specific plans and responsibility for implementation flow "upward" (Figure 7).

The "resources-down, plans-up" relationship has recently become the subject of an Office of Management and Budget (OMB) study (Bureau of National Affairs, Federal Contracts Report, 1978). "Cooperative agreements" (as defined by PL 95-224, Federal Grant

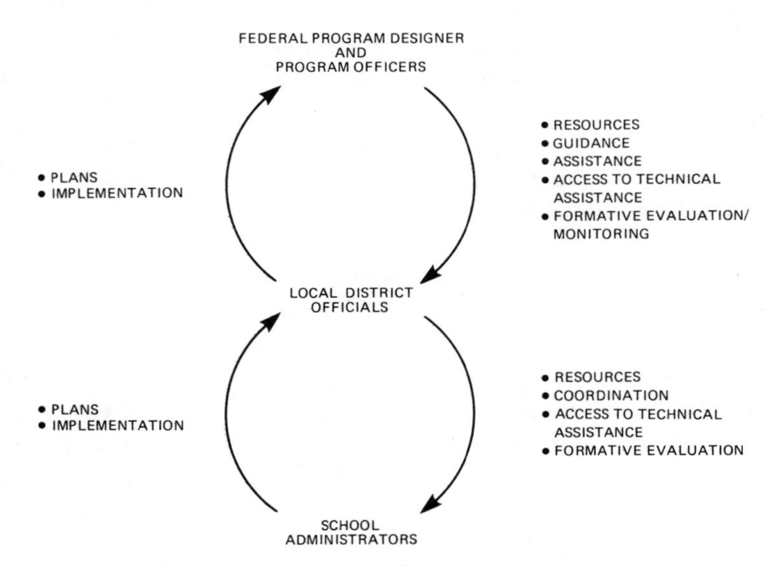

Figure 7. The federal-local relationship in federally funded change efforts.

and Cooperative Agreement Act of 1977, and discussed in the Federal Contracts Report No. 742, 8/21/78) are essentially financial-assistance relationships—grants—with explicit technical-assistance and monitoring responsibilities at the federal level. Although not widely in use at the present, cooperative agreements can facilitate both flexibility and control in federal-local relationships. As a consequence, future programs using this mechanism may avoid the problems faced by the 10 ES districts, which held a formal contractual relationship with ES/Washington which was both legally and psychologically much more binding. Future studies of change could build upon the findings of this study by utilizing this assistance mechanism as the basis of the federal-local relationship.

The Viability of Comprehensive Change

Comprehensive change—the theme of the ES program—may be viewed as a goal or as a strategy for change. If we view comprehensiveness as a program *goal,* it may have been an inappropriate one because it represents an ideal of change which could probably not be met. Most schools and school districts found it difficult to devote most of their energies and resources to the pursuit of radically different educational activities. Although we did not find any evidence that the school districts participating in the Rural ES program had been "transformed," we have been able to show that change did take place, although it varied across districts and among schools within districts. On the other hand, comprehensiveness as a strategy rather than a goal—that is, a deliberate attack on many fronts at the same time—seems to be viable.

"Comprehensiveness" implies planning and coordination of efforts to modify the constituent elements of an educational system; it also implies a concern for both the long-term and short-term effects of intervention within and across schools, districts, and communities. However grandiose or naïve its initial goals, the ES program has shown the utility and necessity of planned, coordinated, cooperative efforts to improve educational quality and opportunity. In sum, comprehensive change may not be achievable, but the use of comprehensive strategy in planning and implementing change holds promise.

References

Abt, W., & Magidson, J. Reforming schools: Problems in program implementation and evaluation. Beverly Hills, Calif.: Sage Publications, 1980.

Adams, C., Kellogg, R., & Schroeder, R. Decision-making and information systems in colleges: An exploratory study. *Journal of Higher Education*, 1976, *47*(1), 33–49.

Alderfer, C. Effect of individual groups and intergroup relations on attitudes toward a management development program. *Journal of Applied Psychology*, 1971, *55*, 301–311.

Anderson, J. Bureaucratic rules: Barriers of organizational authority. *Educational Administration Quarterly*, 1966, *2*, 7–34.

Argyris, C. Understanding human behavior in organizations: One viewpoint. In M. Haire (Ed.), *Modern organization theory*. New York: Wiley, 1959.

Argyris, C. *The applicability of organizational sociology*. Cambridge: Cambridge University Press, 1972.

Arnn, J., & Strickland, B. Human considerations in the effectiveness of system approaches. *Educational Technology*, 1975, *15*(8), 13–17.

Bailey, S., & Mosher, E. *ESEA: The Office of Education administers a law*. Syracuse: Syracuse University Press, 1968.

Baldridge, J. *The analysis of organizational change: A human relations strategy versus a political systems strategy*. Stanford, Calif.: Stanford University Press, 1971.

Baldridge, J., & Burnham, R. Organizational innovation: Individual, organizational and environmental impacts. *Administrative Science Quarterly*, 1975, *20*, 165–176.

Baldridge, J., & Deal, T. (Eds.). *Managing change in educational organizations*. Berkeley, Calif.: McCutchan, 1975.

Barnard, C. *The functions of the executive*. Cambridge: Harvard University Press, 1938.

Bennis, W. *Changing organizations: Essays on the development and evolution of human organization*. New York: McGraw-Hill, 1966.

Bennis, W., Benne, K., & Chin, R. (Eds.). The planning of change (2nd ed.). New York: Holt, Rinehart & Winston, 1969.

Bentzen, M. *Changing schools: The magic feather principle*. New York: McGraw-Hill, 1974.

Berelson, B., & Steiner, G. *Human behavior: An inventory of findings*. New York: Harcourt, Brace and World, 1964.

Berke, J., & Kirst, M. *Federal aid to education: Who benefits? Who governs?* Lexington, Mass.: Lexington Books, 1972.

Berman, P., & McLaughlin, M. *Federal programs supporting educational change.* Vol. 1. *A model of educational change.* Santa Monica, Calif.: Rand Corporation, 1974.

Berman, P., & McLaughlin, M. *Federal programs supporting educational change.* Vol. 4. *The findings in review.* Santa Monica, Calif.: Rand Corporation, 1975.

Berman, P., & McLaughlin, M. *Federal programs supporting educational change.* Vol. 7. *Factors affecting implementation and continuation.* Santa Monica, Calif.: Rand Corporation, 1977.

Berman, P., & McLaughlin, M. *Federal programs supporting educational change.* Vol. 8. *Implementing and sustaining innovation.* Santa Monica, Calif.: Rand Corporation, 1978.

Berman, P., & Pauly, E. *Federal programs supporting change.* Vol. 2. *Factors affecting change agent projects.* Santa Monica, Calif.: Rand Corporation, 1975.

Bidwell, C. The school as a formal organization. In J. March (Ed.), *Handbook of organizations.* Skokie, Ill.: Rand McNally, 1965.

Bishop, D. The role of the local administrator in reorganizing elementary schools to test a semi-departmentalized plan. *Journal of Educational Sociology,* 1961, *34,* 344–348.

Blau, P. *The dynamics of bureacracy.* Chicago: University of Chicago Press, 1955.

Blau, P. Interdependence and hierarchy in organizations. *Social Science Research,* 1972, *1,* 1–24.

Blau, P., & Schoenherr, R. *The structure of organizations.* New York: Basic Books, 1971.

Bons, P., & Fiedler, F. Changes in organizational leadership and the behavior of relationship and task motivated leaders. *Administrative Science Quarterly,* 1976, *21,* 453–473.

Bowers, D. OD techniques and their results in 23 organizations. *Journal of Applied Behavioral Science,* 1973, *9,* 21–43.

Bowles, S., & Gintis, H. *IQ in the United States class structure.* Cambridge: Harvard University Press, 1972.

Bracht, F. "Experimental factors related to aptitude-treatment interactions." *Review of Educational Research,* 1970, *40,* 627–646.

Brickell, H. State organization for educational change: A case study and proposal. In M. Miles (Ed.), *Innovation in education.* New York: Teachers College, Columbia University, 1964.

Buckley, W. *Sociology and modern systems theory.* Englewood Cliffs, N.J.: Prentice-Hall, 1967.

Budding, D. *Draft discussion paper on experimental schools program.* 1972. (Original carries no title, author, or date. Facts of publication established in conversation with the author.)

Burns, A. *From rural schools project to rural schools problem.* Cambridge, Mass.: Abt Associates, May, 1979.

Burns, T., & Stalker, G. *The management of innovation.* London: Tavistock, 1961.

Caplow, T. *Principles of organization.* New York: Harcourt, Brace and World, 1964.

Carlson, R. *Adoption of educational innovations.* Eugene, Oregon: Center for the Advanced Study of Educational Administration, 1965.

Carlson, R. Summary and critique of educational diffusion research. *Research implications for educational diffusion.* Lansing: Michigan Department of Education, 1968.

Charters, W., & Jones, J. On neglect of the independent variable in program evaluation. In J. Baldridge & T. Deal (Eds.), *Managing change in educational organizations*. Berkeley, Calif.: McCutchan, 1975.

Charters, W., & Pellegrin, R. Barriers to the innovation process: Four case studies of differentiated staffing. *Educational Administration Quarterly*, 1972, *9*, 77–81.

Clark, B. *The open-door college: A case study*. New York: McGraw-Hill, 1960.

Clark, B. The organizational saga in higher education. *Administrative Science Quarterly*, 1972, *2*(1), 78–104.

Clinton, C. *Local success and federal failure: A study of community development and change in the rural south*. Cambridge, Mass.: Abt Books, 1979. (a)

Clinton, C. Shiloh County: A matter of agendas. In R. Herriott & N. Gross (Eds.), *The dynamics of planned educational change*. Berkeley, Calif.: McCutchan, 1979. (b)

Coch, L., & French, J. Overcoming resistance to change. *Human Relations*, 1971, *1*, 512–532.

Cohen, E., Deal, T., Meyer, J., & Scott, W.R., Technology and teaming in the elementary school. *Sociology of Education*, 1979, *52*, 20–33.

Colfer, A., & Colfer, C. *Becoming American: Life and learning in an American village*. Cambridge, Mass.: Abt Associates, May, 1979.

Cooke, R. (1972). Personal communications, as cited in G. Zaltman, R. Duncan, & J. Holbek (Eds.), *Innovation and organizations*. New York: Wiley, 1973.

Corwin, R. *Militant professionalism: A study of bureaucratic conflict in high schools*. New York: Appleton, Century, Crofts, 1970.

Corwin, R. Strategies for organizational change. *American Sociological Review*, 1972, *37*, 441–454.

Corwin, R. *Reform and organizational survival: The teacher corps as an instrument of educational change*. New York: Wiley, 1973.

Corwin, R. *Patterns of federal-local relationships in education: A case study of the Rural Experimental Schools program*. Cambridge, Mass.: Abt Associates, 1977.

Coser, L. *The functions of social conflict*. Glencoe, Ill.: Free Press, 1956.

Coughlan, R., & Zaltman, G. *Implementing the change agent team concept*. Paper presented at annual meeting of American Educational Research Association, Chicago, April, 1972.

Coughlan, R., Cooke, R., & Safer, L., Jr. *An assessment of a survey feedback–problem solving–collective decision intervention in schools*. Final Report, U.S. Office of Education, Small Grants Division, Area V, Project No. OE-105, 1972.

Cyert, R., & March, J. *A behavioral theory of the firm*. Englewood Cliffs, N.J.: Prentice-Hall, 1963.

Deal, T., & Celotti, L. How much influence do and can educational administrators have on classrooms. *Phi Delta Kappa*, 1980, *61*(7), 471–474.

Deal, T., Meyer, J., & Scott, W. Organizational influences on educational innovations. In J. Baldridge & T. Deal (Eds.), *Managing change in educational organizations*. Berkeley, Calif.: McCutchan, 1975.

Donnelly, W. *Continuity and change in a rural school district: Constantine, Michigan*. Cambridge, Mass.: Abt Associates, May, 1979.

Donnelly, W. Arcadia: Local initiatives and adaptation. In R. Herriott & N. Gross (Eds.), *The dynamics of planned educational change*. Berkeley, Calif.: McCutchan, 1979.

Downs, G. Jr., & Mohr, L.B. Conceptual issues in the study of innovation. *Administrative Science Quarterly*, 1976, *21*, 700–714.

Doyle, W., Crist-Whitzel, J., Donicht, T., Eixenberger, D., Everhart, R., McGeever, J.,

Pierce, D., & Toepper, R. *The birth, nurturance, and transformation of an educational reform*. Portland, Oregon: Northwest Regional Educational Laboratory, 1976.
Dunn, W., & Swierzcik, F. Planned organizational change: Towards a grounded theory. *Journal of Applied Behavioral Science*, 1977, *13*(2), 135–158.
Emrick, J. *Evaluation of the national diffusion network*. Menlo Park: Calif.: Stanford Research Institute, 1977.
Etzioni, A. *Modern organizations*. Englewood Cliffs, N.J.: Prentice-Hall, 1964.
Fantini, M. Community participation: Many faces, many directions. *Educational Leadership*, 1972, *29*, 674–680.
Fiedler, F. The effect of leadership training and experience: A contingency model and interpretation. *Administrative Science Quarterly*, 1972, *17*, 453–470.
Firestone, W. Butte-Angels Camp: Conflict and transformation. In R. Herriott & N. Gross (Eds.), *The dynamics of planned educational change*. Berkeley, Calif.: McCutchan, 1979.
Firestone, W. *Great expectations for small schools: The limitations of federal projects*. New York: Praeger, 1980.
Fitzsimmons, S., & Freedman, A. *Rural community development: A model for programs, planning, and research*. Cambridge, Mass.: Abt Books, 1980.
Fitzsimmons, S., Wolff, P., & Freedman, A. (Eds.), *Rural America: A social and educational history of ten communities* (2 vols.). Cambridge, Mass.: Abt Associates, 1975.
Ford Foundation. *A foundation goes to school*. New York: Ford Foundation, 1972.
Forehand, G., Ragosta, M., & Rock, D. *Conditions and processes of effective school desegregation*. Princeton, N.J.: Educational Testing Service, 1976.
French, J., Israel, J., & Dagfinn, D. An experiment on participation in a Norwegian factory. *Human Relations*, 1960, *13*, 3–19.
Friedlander, F., & Brown, L. Organization development. *Annual Review of Psychology*, 1974, *25*, 313–341.
Fullan, M., & Pomfret, A. Review of research on curriculum and instruction implementation. *Review of Educational Research*, 1977, *47*, 335–397.
Gaynor, A. *The study of change in educational organizations: A review of the literature*. Paper presented at University Council for Educational Administration (UCEA), Ohio State University Career Development Seminar, Columbus, Ohio, May 1975.
Gephart, W. *Problems in measuring the degree of implementation of an organization*. Paper presented at annual meeting of American Educational Research Association, San Francisco, April 1976.
Giacquinta, J. The process of organizational change in schools. In F. Kerlinger (Ed.), *Review of research in education*. Itaska, Ill.: F.E. Peacock, 1973.
Gideonse, H. Designing federal policies and programs to facilitate local change efforts. In R. Herriott & N. Gross (Eds.), *The dynamics of planned educational change*. Berkeley, Calif.: McCutchan, 1979.
Glaser, E., & Ross, H. *Increasing utilization of applied research results*. Los Angeles: Human Interaction Research Institute, 1971.
Goodlad, J., & Anderson, R. *The nongraded elementary school*. New York: Harcourt, Brace and World, 1963.
Gouldner, A. Organizational analysis. In R. Merton, L. Bloom, & L. Cottrell, Jr. (Eds.), *Sociology Today*. New York: Basic Books, 1959.
Greenwood, P., Mann, D., & McLaughlin, M. *Federal programs supporting educational change*. Vol. 3. *The process of change*. Santa Monica, Calif.: Rand Corporation, 1975.
Griffiths, D. Systems theory and school districts. In P. Sexton (Ed.), *Readings on the school in society*. Englewood Cliffs, N.J.: Prentice-Hall, 1967.
Gross, N. Basic issues in the management of educational change efforts. In R. Herriott &

N. Gross (Eds.), *The dynamics of planned educational change.* Berkeley, Calif.: McCutchan, 1979.

Gross, N., & Herriott, R. Conclusions and implications. In R. Herriott & N. Gross (Eds.), *The dynamics of planned educational change.* Berkeley, Calif.: McCutchan, 1979.

Gross, N., Giacquinta, J., & Bernstein, M. *Implementing organizational innovations.* New York: Basic Books, 1971.

Grusky, O. Corporate size bureaucratization and managerial succession. *American Journal of Sociology,* 1961, *67,* 261–269.

Guest, R. *Organizational change: The effect of successful leadership.* Homewood, Ill.: Dorsey, 1962.

Hackman, J., & Lawler, E. Employee reactions to job characteristics. *Journal of Applied Psychology,* 1971, *55,* 259–286.

Hage, J., & Aiken, M. Program change and organizational properties: A comparative analysis. *American Journal of Sociology,* 1967, *72,* 503–519.

Hage, J., & Aiken, M. *Social change in complex organizations.* New York: Random House, 1970.

Hage, J., & Dewar, R. Elite values versus organizational structure in predicting innovation. *Administrative Science Quarterly,* 1973, *18,* 279–290.

Hall, G., & Loucks, S. *A developmental model for determining whether or not the treatment really is implemented.* Austin: Research and Development Center for Teacher Education, University of Texas at Austin, 1976.

Hall, G., Loucks, S., Rutherford, W., & Newlove, B. Levels of use of the innovation: A framework for analyzing innovation adoption. *Journal of Teacher Education,* 1975, *26* (1), 52–56.

Hall, R. *Organizations: Structure and process.* Englewood Cliffs, N.J.: Prentice-Hall, 1972.

Hall, R., Haas, J., & Johnson, N. Organizational size, complexity and formalization. *American Sociological Review,* 1967, *32,* 903–911.

Havelock, R., Huber, R., & Zimmerman, S. *Major works on change in education: An annotated bibliography.* Ann Arbor: University of Michigan Press, 1969.

Havelock, R., Guskin, A., Frohman, M., Havelock, M., Hill, M., & Huber, J. *Planning for innovation through dissemination and utilization of knowledge.* Ann Arbor, Mich.: Institute for Social Research, 1969.

Heathers, G. *Organizing schools through the dual progress plan.* Danville, Ill.: Interstate, 1967.

Hennigh, L. *Cooperation and conflict in long-term educational change.* Cambridge, Mass.: Abt Associates, May, 1979.

Herriott, R. The federal context: Planning, funding and monitoring. In R. Herriott & N. Gross (Eds.), *The dynamics of planned educational change.* Berkeley, Calif.: McCutchan, 1979.

Herriott, R. *Federal initiatives and rural school improvement: Findings from the Experimental Schools Program.* Cambridge, Mass.: Abt Associates, March 1980.

Herriott, R., & Gross, N. (Eds.). *The dynamics of planned educational change.* Berkeley, Calif.: McCutchan, 1979.

Herriott, R., & Hodgkins, B. *The environment of schooling.* Englewood Cliffs, N.J.: Prentice-Hall, 1973.

Herriott, R., & Rosenblum, S. *First steps towards planned change.* Cambridge, Mass.: Abt Associates, 1976.

Horner, M., & Walsh, M. Psychological barriers to success in women. In R. Knudsin (Ed.), *Women and success: The anatomy of achievement.* New York: William Morrow, 1974.

Hyman, H., Wright, C., & Hopkins, T. *Applications of methods of evaluations: Four studies of the encampment for citizenship*. Berkeley & Los Angeles: University of California Press, 1962.

Iannaccone, L., & Lutz, F. *Politics, power and policy: The governing of a school district.* Columbus, Ohio: Charles E. Merrill, 1970.

Jastrzab, J., Louis, K., & Rosenblum, S. *Memorandum on clustering the structure and culture variables in the organizational change study.* Cambridge, Mass.: Abt Associates, 1977.

Katz, D., & Kahn, R. *Social psychology of organizations.* New York: Wiley, 1966.

Katz, M. *Class, bureaucracy and schools.* New York: Praeger, 1971.

Kent, J. The management of educational change efforts in school systems. In R. Herriott & N. Gross (Eds.), *The dynamics of planned educational change.* Berkeley, Calif.: McCutchan, 1979.

Keys, C.B., & Bartunek, J. Organization development in schools: Goal agreement, process skills and diffusion of change. *Journal of Applied Behavioral Science,* 1979, *15,* 61–78.

Kimberly, J. Organizational size and the structuralist perspective: A review, critique, and proposal. *Administrative Science Quarterly,* 1976, *21,* 571–597.

Kirst, M. The growth of federal influence in education. In C. Gordon (Ed.), *The uses of the sociology of education.* Chicago: University of Chicago Press, 1973.

Kirst, M. Strengthening federal-local relationships supporting educational change. In R. Herriott & N. Gross (Eds.), *The dynamics of planned educational change.* Berkeley, Calif.: McCutchan, 1979.

Klatsky, S. Relationship of organizational size to complexity and coordination. *Administrative Science Quarterly,* 1970, *15,* 248–438.

Klonglan, G., & Coward, W., Jr. The concept of symbolic adoption: A suggested interpretation. *Rural Sociology,* 1970, *35,* 77–83.

Lawrence, P., & Lorsch, J. *Organization and environment: Managing differentiation and integration.* Boston: Harvard Business School, Division of Research, 1967.

Likert, R. *The human organization: Its management and value.* New York: McGraw-Hill, 1967.

Lippitt, R. Consultation: Traps and potentialities. In R. Herriott & N. Gross (Eds.), *The dynamics of planned educational change.* Berkeley, Calif.: McCutchan, 1979.

Lischeron, J., & Wall, T. Employee participation: An experimental field study. *Human Relations,* 1975, *28,* 863–884.

Litwak, E. Models of bureaucracy which permit conflict. *American Journal of Sociology,* 1961, *69,* 177–185.

Litwak, E., & Rothman, J. Toward the theory and practice of coordination between formal organizations. In W. Rosengren & M. Lefton (Eds.), *Organizations and clients.* Columbus, Ohio: Charles E. Merrill, 1977.

Lortie, D. *Schoolteacher: A sociological study.* Chicago: University of Chicago Press, 1975.

Louis, K. Dissemination of information from bureaucracies to local schools: The role of the linking agent. *Human Relations,* 1977, *30,* 25–42.

Louis, K., & Rosenblum, S. *Participation and the implementation of planned change.* Paper presented at annual meeting of American Educational Research Association, New York, April 1977.

Louis, K., & Sieber, S. *The dispersed organization: A comparative study of an educational extension program.* Norwood, N.J.: Ablex, 1979.

Louis, K., Molitor, J., & Rosenblum, S. *System change, system linkage and program implementation.* Paper presented at annual meeting of American Educational Research Association, San Francisco, April 1979.

Louis, K., Rosenblum, S., & Molitor, J. *Linking R&D with local schools: Strategies for school improvement.* Cambridge, Mass.: Abt Associates, 1981.

Mann, D. Making change happen? *Teachers College Record,* 1976, *77*(3), 313–322.

Mann, F., & Neff, F. *Managing major change in organizations.* Ann Arbor, Mich.: Foundation for Research on Human Behavior, 1961.

March, J., & Olsen, N. *Ambiguity and organizational choice.* Bergen, Norway: Universitetsforlaget, 1976.

March, J., & Simon, H. *Organizations.* New York: Wiley, 1958.

Mechanic, D. Sources of power of participants in complex organizations. *Administrative Science Quarterly,* 1962, *7,* 349–364.

Messerschmidt, D. The local-federal interface in rural school improvement. Cambridge, Mass.: Abt Associates, May, 1979.

Messerschmidt, D. River District: A search for unity admidst diversity. In R. Herriott & N. Gross (Eds.), *The dynamics of planned educational change.* Berkeley, Calif.: McCutchan, 1979.

Meyer, M. Organizational domains. *American Sociological Review,* 1975, *37,* 434–440.

Meyer, J., & Rowan, B. Institutionalized organizations: Formal structure as myth and ceremony. *American Journal of Sociology,* 1977, *83,* 340–363.

Miles, R. *Theories of management: Implications for organizational behavior and development.* New York: McGraw-Hill, 1975.

Milo, N. Health care organizations and innovations. *Journal of Health and Social Behavior,* 1971, *12,* 163–173.

Milstein, M. *Impact and response: Federal aid and state education agencies.* New York: Teachers College Press, 1977.

Morse, N., & Reimer, E. The experimental change of a major organizational variable. *Journal of Abnormal and Social Psychology,* 1956, *52,* 120–129.

National Institute of Education. *NIE: Its history and programs.* Washington: National Institute of Education, 1974.

Naumann-Etienne, M. *Bringing about open education: Strategies for innovation.* Unpublished doctoral dissertation, University of Michigan, 1974.

Oettinger, A., & Marks, S. Educational technology: New myths and old realities. In Y. Hasenfeld & R. English (Eds.), *Human service organizations.* Ann Arbor: University of Michigan Press, 1974.

Owens, R. *Organizational behavior in schools.* Englewood Cliffs, N.J.: Prentice-Hall, 1970.

Parsons, T. *The social system.* Glencoe, Ill.: Free Press, 1951.

Parsons, T. *Structure and process in modern society.* Glencoe, Ill.: Free Press, 1960.

Pennings, J. Dimensions of organizational influence and their effectiveness correlates. *Administrative Science Quarterly,* 1976, *21,* 688–699.

Perrow, C. *Complex organizations: A critical essay.* Glenview, Ill.: Scott, Foresman, 1972.

Perrow, C. The analysis of goals in complex organizations. In Y. Hasenfeld & R. English (Eds.), *Human service organizations.* Ann Arbor: University of Michigan Press, 1974.

Peterson, V.T. *Policy implementation as a loosely coupled organizational adaptation process.* Paper presented at annual meeting of American Educational Research Association, New York, April 1977.

Pincus, J. Incentives for innovation in the public schools. *Review of Educational Research,* 1974, *44,* 113–144.

Price, J. *Handbook of organizational measurement.* Lexington, Mass.: D.C. Heath, 1972.

Pritchard, R., & Karasick, B. The effect of organizational climate on management, job performance and job satisfaction. *Organizational Review,* 1973.

Pugh, D., Hickson, D., Hinings, C., & Turner, C. Dimensions of organizational structure. *Administrative Science Quarterly,* 1968, *13,* 65–106.

Reagen, M. *The new federalism.* New York: Oxford University Press, 1972.

Rogers, E. *Diffusion of innovations.* New York: Free Press, 1962.

Rogers, E., & Shoemaker, F. *Communication of innovations.* New York: Free Press, 1971.

Rosenblum, S., & Louis, K. *A measure of change: Implementation in ten rural school districts.* Cambridge, Mass.: Abt Associates, 1977.

Royster, E., Baltzell, D., & Simmons, F. *Study of the Emergency School Aid Act Magnet School Program.* Cambridge, Mass.: Abt Associates, 1979.

Sapolsky, H. Organizational structure and innovation. *Journal of Business,* 1967, *40,* 497–510.

Sarason, S. *The culture of schools and the problem of change.* Boston: Allyn & Bacon, 1971.

Schein, E. *Process consultation: Its role in organization development.* Reading, Mass.: Addison-Wesley, 1969.

Selznick, P. *Leadership in administration.* Evanston, Ill.: Row, Peterson & Co., 1957.

Shepard, H. Innovation-resisting and innovation-producing organizations. *Journal of Business,* 1967, *40*(4), 470–477.

Sieber, S. Images of the practitioner and strategies of educational change. *Sociology of Education,* 1972, *45,* 365–382.

Sieber, S. The integration of fieldwork and survey methods. *American Journal of Sociology,* 1973, *78,* 1335–1359.

Sieber, S. Federal support for research and development in education and its effects. In *Seventy-third yearbook of the National Society for the Study of Education.* Chicago: National Society for the Study of Education, 1974.

Siegal, B. Models for the analysis of the educative process in American communities. In G. Spindler (Ed.), *Education and anthropology.* Stanford, Calif.: Stanford University Press, 1955.

Simon, H. *A study of decision-making processes in administrative organizations* (2nd ed.). New York: Free Press, 1965.

Smith, L., & Keith, P. *Anatomy of educational innovation.* New York: Wiley, 1971.

Smith, T., & Zopf, P. *Principles of inductive rural sociology.* Philadelphia: F.A. Davis, 1970.

Sproull, L., Weiner, S., & Wolf, D. *Organizing an anarchy: Belief, bureaucracy and politics in the National Institute of Education.* Chicago: University of Chicago Press, 1978.

Stannard, C. *Problems of direction and coordination.* Cambridge, Mass.: Abt Associates, May, 1979.

Tannenbaum, A. *Control in organizations.* New York: McGraw-Hill, 1960.

Terreberry, S. The evolution of administrative environments. *Administrative Science Quarterly,* 1968, *12,* 590–613.

Thompson, J. *Organizations in action.* New York: McGraw-Hill, 1967.

Tyack, D. *The one best system.* Cambridge: Harvard University Press, 1974.

U.S. Congress, Public Law 92-318.

U.S. Department of Labor. *Monthly Labor Review.* August 1974.

U.S. Senate, Subcommittee on Rural Development. *HEW programs for rural America.* Washington, D.C.: U.S. Government Printing Office, 1975, pp. 36–61.

Vall, M., Bolas, C., & Kang, T. Applied social research in industrial organizations: An evaluation of functions, theory and methods. *Journal of Applied Behavioral Science,* 1976, *12*(2), 158–177.

Vanecko, J., Archambault, F., & Ames, N. *ESEA Title I allocation policy: Demonstration study—results of first-year implementation.* Cambridge, Mass.: Abt Associates, 1978.

Wacaster, C. Jackson County: Local norms, federal initiatives and administrator performance. In R. Herriott & N. Gross (Eds.), *The dynamics of planned educational change.* Berkeley, Calif.: McCutchan, 1979.

Watson, G. *Concepts for social change.* Washington, D.C.: Cooperative Project for Educational Development, National Institute for Applied Behavioral Science, 1967.

Wayland, S. Structural features of American education on basic features in innovation. In M. Miles (Ed.), *Innovation in education.* New York: Teachers College, Columbia University, 1964, pp. 587–613.

Webb, E., Campbell, D., Schwartz, R., & Sechrest, L. *Unobtrusive measures.* Chicago: Rand McNally, 1966.

Weber, M. *The theory of social and economic organizations.* Glencoe, Ill.: Free Press, 1947.

Weick, K. Educational organizations as loosely coupled systems. *Administrative Science Quarterly,* 1976, *21,* 1–9.

Wilson, J. Innovation in organization: Notes toward a theory. In J.D. Thompson (Ed.), *Approaches to organizational design.* Pittsburgh: University of Pittsburgh Press, 1966, pp. 193–218.

Woodward, J. *Industrial organization.* London: Oxford University Press, 1965.

Yin, R. Production efficiency versus bureaucratic self-interest: Two innovation processes? *Policy Sciences,* 1977, *8,* 381–399.

Yin, R., & Quick, B. *Thinking about routination.* Santa Monica, Calif.: Rand Corporation, 1977.

Yin, R., Heald, K., & Vogel, M. *Tinkering with the system.* Lexington, Mass.: Lexington Books, 1977.

Zaltman, G., Duncan, R., & Holbek, J. (Eds.). *Innovation and organizations.* New York: Wiley, 1973.

APPENDIX A

Site-by-Site Descriptions of the Ten ES Projects as Implemented and Continued

SALMON POINT

Population (1970)	*272*
Number of schools	*2*
Number of pupils (1973–1974)	*146*
Amount of ES planning grant (1972–1973)	*$90,500*
Amount of ES implementation funding (1973–1976)	*$265,365*

In planning its ES project, Salmon Point continued and expanded innovative efforts previously under way. Salmon Point's planners focused on the district's extreme isolation and small size. A district chronically afflicted with high seasonal unemployment, Salmon Point's project emphasized a curriculum featuring individualized instruction and offering vocational options both for those who planned to remain in the area and those who planned to leave Salmon Point (approximately 70% each year) for further training and education.

During the first year of implementation (1973–1974), Salmon Point put into operation several components designed to individualize and expand its curriculum for all pupils from preschool through high school. Over 60% of the ES resources went into continuing the previously instituted "Basic School." For high school students a "Career School" alternative to traditional school allowed students to forego the classroom entirely. Students who participated in this component had to fulfill 53 practical living requirements—including skills such as academic knowledge, first aid, and learning to fill out an income tax form. Although a child in the Career School had no formal daily commitments, the child selected an advisor who monitored progress in the practical living requirements.

"Conventional School" provided an innovative approach to a set of traditional curriculum options for students who did not opt for the Career School. A "Career Resource Center" was established to give junior and senior high-school students and teachers materials on educational and career opportunities.

For children below school age, an "Early School" component was set up to teach preschoolers reading readiness and word-recognition skills. An "In Home" program for three- to four-year-olds allowed parents to use preschool educational material and satellite television programming.

For all students above third grade, a very extensive and far-reaching "Student Travel" component was offered on an individual basis for academic credit with the goal of preparing them for life outside the isolated Salmon Point community. Several of the trips included international travel. Films and written materials were used as preparatory materials before departure. Community residents participated in fund-raising activities with students (travel was funded approximately 50% by ES) and as travel escorts. Students participated (depending on age) in developing their own itinerary and in making travel arrangements.

For adults in the community, many of them dropouts or apathetic toward schooling, a "Community Education" component provided educational and recreational opportunities, intended to reawaken interest in education.

In addition to the curriculum innovations noted above, other ES components which were implemented were: a system of "Accountability" in which teachers were taught to write precise objectives for each child to be shared with the student's parents and the administration. These were discussed with parents at in-home conferences in which parents were encouraged to add any special skills they wished their children to learn. "In-Service Training" was held for teachers for individualized math and reading instruction through packaged learning programs.

Of the 12 components whose implementation was begun in the fall of 1973, five were discontinued in the following year. These were the "Career School," component, which at the end of the first year retained only one of the eight students originally enrolled, the "Teacher Exchange" component, "Information Study" component, the "Community Involvement" component, and the "Career Resources Center" component. However, "Career Day" (two days per year), in which community members came into the school to describe their work, survived. In addition, four components identified in the ES project plan were either never implemented or discontinued within the first year: "Extension of the School Day and Year," "Planned Program Budget System," "Learning Counselors" (older students tutoring younger ones), and "Community Development."

Only five components of the original ES project survived by the end of third school year (1975–1976): "Basic School," "Conventional School," "Community Education," "Student Travel," and "Accountability."

Salmon Point ranked "moderate" on continuation. Although the district remains one of the most innovative and least traditional of the ES districts, many of these changes were begun prior to the ES project. However, they became more entrenched during the project years. For example, the elementary school is completely organized by teaching specialists, resembling a departmentalized secondary school rather than having one teacher for all subjects at a particular grade level. The high school is organized through a series of minicourses rather than on a two-semester basis, and this arrangement has been formalized during the ES years. The single greatest success which continues is the very extensive travel program. This program, now supported through massive community and school involvement in fund raising, has had academic spin-off effects as well, through the imposition of eligibility criteria (academic achievement) for students to participate.

PRAIRIE MILLS

Population (1970)	*5,038*
Number of schools	*4*
Number of pupils (1973–1974)	*1,659*
Amount of ES planning grant (1972–1973)	*$99,250*
Amount of ES implementation funding (1973–1976)	*$715,611*

This Midwestern community faced chronic problems of a high outmigration rate of its young people, a large percentage of high school dropouts, and a lack of interest in continued education after high school. In planning an ES project school officials addressed these and other concerns, among them poor communication between the schools and the community and a lack of community participation in the schools.

Even during the planning year Prairie Mills began "Evaluation" through a series of periodic questionnaires administered to students and faculty members to assess existing program effects and relay the information to decision makers.

In the fall of 1973, an "Early Childhood Education" program was begun to identify learning disabilities through perceptual testing of preschool children and provide training to parents to work toward solutions of identified problems.

Components dealing with individualization of the curriculum began early in the 1973–1974 school year. Primary among these were packaged programs in "Math" and "Language Arts" for both elementary and secondary students. A

"Careers" component provided instruction about career possibilities, and for high-school students included a curriculum in office practice. "Educational Travel" for students broadened the curriculum and supplemented classroom instruction.

Prairie Mills planned components which would enhance the sense of opportunity of living a satisfying life in a rural area (particularly through outdoor activities). During the 1973–1974 school year, Prairie Mills began an ambitious program of "Physical Education" and recreation designed to orient students toward lifetime involvement in sports. This program was also used to identify and correct disabilities and handicaps. An "Outdoor Education" component involved students in developing and using an outdoor education facility including primitive camping and a rugged obstacle course. Middle-school students built a pioneer cabin in conjunction with learning community history via the "Community and Family" component. Prairie Mills also set up a "Student Services" component in which older students tutored younger ones through a student services center. A guidance counselor began to work with students in grades 6–9.

Two components, identified as "Improved Staffing" and "Staff Training," gave release time for teachers to work on curriculum development and provided in-service programs to provide all staff with knowledge about a "guidance approach" to teaching.

Improved community relations and community involvement in the schools were facilitated by several components which were implemented in the 1974–1975 school year.

A "Continuing Education" component offered to adults both high-school-completion and enrichment courses. A "Media" program involved community members, teachers, and students in providing media materials and in arranging for printing facilitites in the high school to service school needs.

Also implemented in the second year was Prairie Mills's innovation in "Transportation," the installation on three school buses of television sets turned to educational programs and earphones and tape decks for foreign-language practice—putting travel time to practical use. For the sake of more flexibility in scheduling extracurricular activities, shuttle-bus runs began transporting students outside regularly scheduled runs.

Prairie Mills presents an uneven picture of implementation and discontinuation among its four schools. Although all of the components survived, to a certain extent, into the spring of 1976, they did not survive in all schools in which they had been implemented: "Careers" discontinued in the middle school, "Community and Family" discontinued in the elementary schools and was never begun in the high school, "Student Services" discontinued in the high school.

Although Prairie Mills was one of the highest implementers of the 10 ES districts, it ranked "moderate" on continuation. The district was faced with financial difficulties in the local economy during the later project years, and could not pick up the funding of some of its special programs, such as its

"Travel" program "Shuttle Buses," and "Early Childhood Education." The popular media center continues, supported by other "soft" monies (CETA), which, of course, may be short-lived as well. However, more continuity in the broadened curriculum exists than before, particularly in the middle and high schools. A major innovation in the middle school, the reading laboratory and the teaching of reading as a special subject for all students (frequently not done at the middle-school level) is continuing, as is the guidance program in the middle school. Knowledge that was gained from in-service training in the ES years has recognized value, as demonstrated by the inclusion of "In-Service Day" in the school calendar (which did not exist prior to ES). There is more of a sense of the individual needs of pupils, although it is feared that the heightened awareness and skills of teachers will be lost through the process of staff turnover in the next few years.

BIG SKY

Population 1970	*4,138*
Number of schools	*12*
Number of pupils (1973–1974)	*1,366*
Amount of ES planning grant (1972–1973)	*$46,500*
Amount of ES implementation funding (1973–1976)	*$361,351*

This recently consolidated, far-flung school district was particularly motivated to participate in ES as a catalyst for unity. However, its large size caused administrative difficulties in the coordination of educational services. Its major successes were in the areas of community education and in expanding cultural activities in the school district.

Big Sky, more than any of the other school districts, identified its components in the formal project plan at a level of aggregation corresponding to the facets of educational change. Each of these components was then disaggregated into subcomponent parts. Big Sky was one of few school districts in the program which did not emphasize basic skills in its plans.

In the area of "Curriculum," Big Sky began a two-pronged effort in its schools. These were identified as "Career Education" and "Cultural Education."

The "Career Education" subcomponent began in 1973 for the elementary grades and the following fall for grades 7–12. It was designed to stimulate good career attitudes, awareness, and orientation. "Guidance-Counseling" activities supplemented in-class instruction. In the following year, part of the component at the junior-high level included a series of video programs in career education which originated from a 1974–1975 satellite television project.

The other portion of the curriculum component was directed at "Cultural

Education." Under the auspices of ES, guest poets, artists, and artisans came into the community. Fine arts events, concerts, and a humanities class in three of the four high schools rounded out the effort to make cultural enrichment pervasive through the community and schools.

Big Sky's major innovation in the "Staffing" component was the employment of 3½ FTE "Guidance Counselors," primarily for the district's four high schools but used in K–6 in all schools as well. This new staff was closely involved in the "Career Education" components. Also in conjunction with career education, four "In-Service Workshops" were held for staff at all grade levels.

The communities served by Big Sky's schools were described by the district administrators as "supportive of the schools," and during the ES planning period, citizen "Advisory Committees" played an important part. The "Community Involvement" component was an attempt to continue this active involvement into the implementation period and in evaluation of school system programming. The general trend, however, has been toward less participation by fewer members at a smaller number of meetings.

"Community Involvement" fared far better in its other subcomponent, a program of community–school educational and recreational enrichment classes and activities. A total of 716 adults participated in 48 classes offered in the communities throughout the school district in 1975–1976. The staff, outside of the ES funded director, consisted of community volunteers.

A "Media Center" in Big Sky existed before the ES project, but was expanded and integrated into the "Career Education" component. It provided supportive back-up of media supplies to teachers and classrooms. In concert with another federal program, the center provided television programming which picked up satellite transmission of programs related to career education.

The "Curriculum Education" component was an attempt to achieve district coordination of school programs. It did not get off the drawing board until the summer of 1976, when the state mandated an evaluation program for all districts (apart from ES) and the original ES evaluation plan was adapted to satisfy the state requirements.

A low-implementing district, Big Sky also ranked in the lowest group on continuation. Of the components which were implemented in Big Sky, the career and community school courses are continuing in a somewhat revised curriculum. Evaluation will continue in whatever manner is required by the state.

CLAYVILLE

Population (1970)	*7,080*
Number of schools	*5*

Numbers of pupils (1973–1974)	*1,510*
Amount of ES planning grant (1972–1973)	*$118,000*
Amount of ES implementation funding (1973–1976)	*$567,505*

In the 1960s, a large influx of new residents seeking employment in the burgeoning industrial development in the area caused school and community conflict in Clayville: new residents brought in ideas about schooling that were different from those of the older residents. The local ES project pledged itself to increased community participation and heightened community awareness of school problems. The ES project also planned to emphasize a career education component (especially for the noncollege-bound student) based on local opportunities in new industries. Individualized teaching techniques would arouse student interest and couteract apathy and a high dropout rate. Poorly educated adults in the community would be offered practical, vocational, and technical classes.

Several "packaged" components were implemented in the curriculum: a "Reading" program, "Continuous Progress Math," and a "discovery-learning Science" program. For all these areas, in-service "Staff Training" in techniques for using the programs was provided to teachers.

One room in the middle school was staffed by a reading clinician and an aide, supplemented by reading teachers and support services in speech, hearing, and social problems.

The attempt by consultants hired specifically for that purpose to install a new teaching method ("Diagnostic Instruction") was eventually abandoned because of teacher resistance. More successfully introduced into the curriculum was a "Career Education" component for all grades in all schools which emphasized (at various levels of detail) "the world of work."

In all schools, libraries were reorganized as "Media Centers" so that all materials were organized by Library of Congress categories. Materials related to specific content were filed together, for example, filmstrips, records, and books.

"Field Trips" for all schools took students to points of interest in and around the community. A community-school "Cultural Enrichment Committee" brought into the district cultural offerings such as an art show and drama groups. Another community-school group formed the "Historical Society" to collect and highlight the history of the region. An ES funded "Newsletter" was published by the schools which also emphasized historical events.

A "Community Advisory Council" was formed in 1972 to give community input into planning the local ES project and to participate in the governance of the project. The council was abandoned in 1974 and never reconstituted.

At the end of the 1975–1976 school year, the curriculum components involving "Reading" and "Career Education" had survived in whole or in part. Others, found costly or not effective in changing achievement level ("Math" and "Science"), were discontinued. In 1975–1976 there was considerable community support for both the "Cultural Enrichment" and "Historical Society" components, which continued to be in effect. But community participation (in the form of the formal "Community Advisory Council") other than in culture and history has discontinued.

Clayville, which ranked "moderate" on continuation, dropped its overriding component, "Diagnostic Instruction," as a separate entity, but incorporated many of its elements into more traditional subject areas. The curriculum components and extensive materials associated with diagnostic instruction have been integrated to varying degrees in the schools' ongoing curriculum and functioning. Although in-service and evaluation activities have decreased, the career education components continue, in large measure because they are now state mandated. "Special Interest" classes and guidance services continue, and the "Steering Committee" created to monitor the ES program continues to function in order to monitor other state-mandated programs. Although Clayville did not achieve the stated objectives of a school system patterned on diagnostic instruction, it is locally perceived to have achieved many local goals and to be an improved school system.

BUTTE–ANGELS CAMP

Population (1970)	*9,858*
Number of schools	*7*
Number of pupils (1973–1974)	*2,350*
Amount of ES planning grant (1972–1973)	*$46,500*
Amount of ES implementation funding (1973–1976)	*$733,840*

Located in a north-central plains state, Butte–Angels Camp has the largest population of the 10 rural districts. The major educational focus of the formal project plan was on individualized education. The district also emphasized bringing community people into the schools as guest speakers, teacher aides, and workshop leaders.

The immediate and central component put into operation during the 1973–1974 school year was "Diagnosis and Prescription" in which curriculum would be devised to fit each individual child after his personal learning needs had been identified. The program demanded training of teachers in both the techniques and the paper work (which was extensive) accompanying the component.

In two elementary schools, flexible scheduling ("Teaming") was implemented to vary group sizes and time periods on a daily basis. In the classroom an "Aide" program gave teachers extra help. By 1975–1976, 44 full- or part-time aides (36 funded by ES) and two teachers (one funded by ES) were employed in the school system to handle either learning disabilities or the special problems of handicapped children.

The "Counselor Social Worker" component involved attempts to use counselors more efficiently. One counselor (of three) dealt with testing and vocational guidance. The other two concentrated on periodic social counseling and began as liaisons with the community through home visits.

The community took an active role in school governance through a "Curriculum Cabinet" component, a district wide parent–teacher administrative council which reviewed all proposed curriculum changes. "Tri-Parties (TAPS) Councils" at each school had the right to make proposals for change to the "Curriculum Cabinet." Originally legislative, the councils subsequently became geared to informing parents about the school programs.

"Evaluation" was handled by a committee of teachers whose purpose was to apply a locally conceived evaluation design. In the third year of this program, they were taking course work in preparation for evaluating other programs in the system as well.

Five additional components were implemented in the 1974–1975 school year. Still in the area of curriculum but moving beyond the classroom, an outdoor classroom was set up as part of the "Environmental Education" component. Elementary students developed a manual for this classroom and participated in exercises and activities. Junior high-school students used the outdoor classrooms as part of a science course. High-school students took courses emphasizing social aspects of the environment. In the general-curriculum areas, a teacher in the high school was given release time to serve as a "Mathematics Coordinator" for math instruction from K to 12.

An "Alternative Notebook" (now indexed as a card catalog) was set up to list treatments which had been found successful in dealing with specific kinds of learning problems. The audiovisual equipment in the district was also catalogued, so that teachers would know what was available.

All of the components except "Tri-Parties" and "Curriculum Cataloguing" survived through the spring of 1976. However, "Diagnosis and Prescription," whose original purpose was to facilitate individualized instruction, became primarily a system of pupil record keeping with little effect on teaching approaches.

Butte–Angels Camp ranked "moderately low" on continuation, in large measure because of the rapidly diminishing scope of its major component, "Diagnosis and Prescription," during the project years. However, more teachers claim to be using individualized instruction than they did previously, and there appears to be a raised consciousness on the part of teachers toward the needs of pupils as individuals. A major if peripheral benefit of ES to the district was to solidify the merger of two previously independent school districts, which formally occurred in 1971 but became more of a reality through the coordination required by ES.

LIBERTY NOTCH

Population (1970)	*3,816*
Number of schools	*4*
Number of pupils (1973–1974)	*1,217*
Amount of ES planning grant (1972–1973)	*$46,500*
Amount of ES implementation funding (1973–1976)	*$336,953*

Located in northern New England, Liberty Notch suffers from a depressed economy. According to the reports of school officials, its residents (almost 30% foreign-born) are generally dissatisfied with both their community and their schools. In order to be eligible for the ES program, three autonomous school districts previously united only for selected administrative purposes agreed to cooperate as one "unified" school district. Perhaps because of this special arrangement, implementation did not take place evenly in all school districts.

Liberty Notch stressed vocational and career education in its formal plan and hoped to provide both adult education and some psychological services to the community. All of the ES Project components were implemented in the first implementation year (1973–1974).

The major curriculum innovation in the basic skills area was a "Language Arts" components, which involved all grade levels in a continuous, integrated, sequential program of reading, listening, speaking, and writing under the direction of an ES funded coordinator.

A "Career Education" component included all grades levels and was intended to increase vocational skills and career awareness among students. Among the opportunities open to older students was an "Intern" program, in which students were trained in a local hospital, and other work-site experiences in the community.

To promote an awareness of "man's interrelatedness with nature," an "Environmental Education" component began in kindgergarten with field trips and nature walks. Older students took overnight camping trips chaperoned by community members, and a high-school class built a house in the woods with materials donated by local industry. Also included within the component were courses in geology, geography, and meteorology.

The "Adult Education" component was originally based on what educators perceived as the urgent needs of the community, and included language arts, computational skills, business training, and a law course. These courses, however, were unpopular with community people and gave way to courses in leisure and recreational activities which provided community residents with opportunities for social and personal interaction.

One of the original focuses of the project was to set up a "Psychological Service" component with the cooperation of a regional mental-health facility.

In the 1975–1976 year this component was discontinued on a formal basis, but there were some continued auxiliary services to other components ($1,800 of ES funds). All but the "Psychological Services" component continued through the 1975–1976 school year.

A relatively low-implementing district, Liberty Notch ranked "moderately low" on continuation. Although it never had a well-coordinated or well-developed project, it now claims to have achieved a more sequential and coordinated curriculum in language arts in the elementary schools, and is in the process of integrating units developed for career education and environmental education into the regular social studies and science curricula in one school. Further, the very popular vocational education programs in the high school as well as the camping program are also continuing.

MAGNOLIA

Population (1970)	*5,908*
Number of schools	*6*
Number of pupils (1973–1974)	*1,507*
Amount of ES planning grant (1972–1973)	*$91,500*
Amount of ES implementation funding (1973–1976)	*$518,133*

This Southern community had the lowest per-capita income of the 10 Rural ES districts with an accompanying high rate of unemployment. Its population (about one-third black and two-thirds white) was primarily employed in forestry and farming. The school district was plagued by the poverty of the area; pupil performance on national tests was poor and there were no support services (such as guidance counselors) provided. The schools wished to increase communication and involvement with the community and to begin to take advantage of the facilities of two nearby universities. Within its ES Project, Magnolia also hoped to deal with lack of education among adults as well as health problems in the community.

In the fall of 1973, an "Early Childhood" component was implemented, offering an academic and social readiness program for five-year-olds. The "Early Childhood" classes were taught by a teacher and an aide each school day in each of the three elementary schools in the district.

Magnolia's "Career Education" component was an attempt to infuse career education into all grade levels above the third grade, not as a separate subject area but as a part of other curriculum areas. Pamphlets were prepared and distributed to teachers, and each teacher decided individually how to handle the infusion process. Although this component was discontinued in the 1974–1975 school year, it was reimplemented in the following year.

"Staff Development" was incorporated into the project because members of the instructional staff were to be primary actors in both curriculum revisions and the design of new programs. This component, which offered various workshops and classes, was seen as supportive of other components and as a principal means for implementing them.

"Evaluation" was also viewed as having a supportive role. It involved the training of local personnel in competencies appropriate for developing a local evaluation program, and then for the development and implementation by these personnel of such a program for the other ES project components.

A "Management Study" component consisted of studies of such aspects of school operations as finances, personnel utilization, and plant management, primarily with an eye to planning for the future. A university consultant team along with local personnel made recommendations to the school board and system administrators as a result of the study.

An advisory committee of school-system patrons was formed by Magnolia as its "Community Involvement" component. Its role was informational and designed to communicate the efforts of the schools to the community and at the same time allow the community an opportunity to contribute in identifying the unmet needs of students.

During the 1974–1975 school year, a "Competency-Based Instructional Component" was implemented. This component was concerned with most areas of the school system's curriculum. Knowledge, attitudes, and skills expected of each student in each subject area at each grade level were listed by principals and the ES project assistant director. Learning activities and resources to develop the skills and competencies were identified and assessment techniques developed.

During that year, as well, a "Health and Physical Education" component began at each elementary school—planned and coordinated by a physical education coordinator paid by ES funds. However, this component did not last more than that one year.

All components except "Health and Physical Education" were still operating in the spring of 1976.

Magnolia never achieved a high scope of implementation and suffered greatly in the continuation stage, ranking "low," in large part because it lost its most popular component, the "Early Childhood" program. This program was not allowed to continue because of a state law prohibiting the use of county funds to support kindergarten programs. This law was passed after the school board had voted unamimously to continue the program with local funds. However, the district feels that the new knowledge and skills gained from staff development continue to reside in the district—at least for as long as the personnel remain. Magnolia also developed some skills in evaluation techniques which are now being used to fulfill state-mandated accountability requirements.

OYSTER COVE

Population (1970)	*1,999*
Number of schools	*2*
Number of pupils (1973–1974)	*341*
Amount of ES planning grant (1972–1973)	*109,000*
Amount of ES implementation funding (1973–1976)	*$379,773*

The second-smallest of the districts in the study, Oyster Cove has nevertheless gained 11% in population in the last 10 years. The economy is based on logging and fishing, the environment beautiful and unspoiled. Residents were generally satisfied with their quality of life but not with the schools—80% felt that change was needed. The plan for Oyster Cove stressed small-group instruction, programmed individual instruction, and study and understanding of the environment.

More than any other project, Oyster Cove displayed a pattern of gradual implementation of its components over a three-year period. In several instances, components were implemented at the high-school level first and later introduced into the elementary school.

Curriculum components in Oyster Cove generally followed the path of individualized instruction. In the fall of 1973, "SRA" and "Craig Individualized Reading" packages as well as "Revised English" were implemented.

"Career Experience" and "Environmental Education" began in the high school in the 1973–1974 school year. Both components were introduced in the elementary school in relevant forms the following year. "Environmental Education" involved students in a course of study concerned with the surrounding beaches and forests.

"Interim" was a special program planned during semester break to take seven high-school students to a nearby large city weekly. "Field Trips" took students out into the surrounding environment, such as the nearby beaches. "Interim" was discontinued, despite its popularity with students, because of some parental complaints about student drinking.

The "Arts and Crafts" component included an auto mechanics shop which became very popular with the students and was considered realistic preparation for future jobs. Local businesses became involved by providing the engines and auto parts.

The "Evaluation" component included a computerized updating of record keeping, and the adoption of an evaluation model (testing and tracking) for the district.

"Independent Study," implemented in 1974–1975, introduced correspondence courses to provide individualized programs for high-school students.

In the 1975–1976 year, "Fountain Valley Support" was introduced as a tracking system for student achievement of reading skills. "Community Education" was also added in that year.

In the spring of 1976, all but two components,—"Revised English" and "Interim"—were still in operation.

Oyster Cove ranked "high" on continuation relative to the other ES districts. Continuation is particularly evident in the high school, where highly popular, career-oriented class offerings continue, many of which are community based. Another core program, "Independent Study," continues in that high school. The district's one elementary school is also continuing an expanded curriculum and the use of new materials and facilities, including environmental and marine education and the "Model Testing" program, which emphasizes the use of testing results for diagnostic purposes. The benefits of in-service training which began under the auspices of the ES project have resulted in a continued emphasis on in-service, which holds great promise for continued improvement in this small district.

TIMBER RIVER

Population	*8,637*
Number of schools	*5*
Number of pupils (1973–1974)	*2,275*
Amount of ES planning grant (1972–1973)	*$120,400*
Amount of ES implementation funding (1973–1976)	*$992,210*

Timber River, which achieved the highest scope of implementation, is primarily a logging community with a high rate of seasonal unemployment. The unemployment problems were a major focus of the formal project plan and were to be addressed through career education components.

The 40 components in Timber River's ES project are more than double the number included in the project of any other district. They are notable as well because each component is discrete and differentiated. For example, Timber River had five components in the area of cultural education alone.

"Outdoor School," "Nature Trails," and "Cultural Enrichment" took students beyond the classroom walls to various parts of the environment—physical and cultural. "Field Experience" provided them with on-the-job training in local industry. The "Media Center" expanded the library and diversified its materials.

The components implemented in Timber River are often closely related and supportive of one another. For example, there were several attempts to expand the school experience and make it more flexible. In this category was "Junior High Program Reorganization" which shifted junior high schools from an annual to a quarter system and increased the variety of course offerings. "Modified Scheduling" in the high school individualized scheduling for high-school students according to their interests and abilities. "Friday Electives" allowed elementary students to pursue their own curriculum interests during two school hours one day a week. The activities are designed by teachers to augment skills learned in curriculum areas. In the area of curriculum the changes included several learning packages: "Social Studies" ("People and Technology" and "Man: A Course of Study"), the special "Mathematics" program, "Developmental Reading," and "Foreign Language" (German for high-school students).

The need for aid both in equipping the ES project and providing classroom assistance for teachers was recognized in Timber River, which implemented an "Instructional Leaders" component which added full-time personnel to assist principals, and a "Differentiated Staffing" component which brought teacher aides, parent aides, and student aides into the classroom. In addition to serving as aides, students in junior and senior high schools began to teach younger students (for academic credit) in a program of "Peer Counseling."

The goal of community involvement was pursued in several ways, not only by on-the-job training but through courses such as the one on burglary prevention taught by police. Senior citizens served as tutors in Timber River and parents volunteered to teach recreation courses. A committee of 18 residents formed the "Coordinating Council," which was informed of and gave advice to the entire ES project.

One reason Timber River was able to do so much implementing at once was the differentiated nature of the components. Another reason may be the detailed and sophisticated nature of the formal project plan prepared by the district; and in that regard the planning grant of $120,4000 which Timber River received—the largest received by any district—is noteworthy.

Timber River ranked "high" on continuation. The district continues to use curriculum guides that were developed under ES, and is enforcing minimum competencies that were established at the elementary-, middle-, and high-school levels, including minimum competencies that were required for high-school graduation. The high school is continuing career education components that were developed during the ES period (although it has eliminated some because of community resistance or lack of support by teachers and pupils), a freshman orientation period, a personalized, ongoing advisor/student relationship, and a better testing and counseling program. The middle school continues to utilize new materials and equipment as well as the media center that were developed by the project, and also continues its use of a modified schedule with special elective programs. Furthermore, it has found that the additional administrative staff that were employed during the project years have been helpful, and local funding has been acquired to support a half-time assistant principal. In the elementary schools, the implementation of new curriculum elements and a curriculum guide continue, and the district now supports teaching aides in grades 1–3, enabling the continuation of more individualization in the classroom. School entry assessments are continuing, along with the expanded and popular "Resident Outdoor School" program for all fifth-graders in the district. In general, Timber River became a more coordinated school district with increased individualization, and, despite a relatively high amount of dissatisfaction with the program on the part of teachers and residents, the school board increased its budget by 17% in order to continue ES sponsored programs and activities.

DESERT BASIN

Population (1970)	*4,535*
Number of schools	*3*
Number of pupils (1973–1974)	*1,503*
Amount of ES planning grant (1972–1973)	*$46,500*
Amount of ES implementation funding (1973–1976)	*$699,680*

A minority group of Mexican-American heritage constitutes 25% of the population of this southwestern community. Among minority group members, the level of education was low and the level of juvenile delinquency high. To meet the needs of this group, Desert Basin planned early childhood, bilingual, and adult education components, as well as a counseling and guidance component. As the largest employer in the community, the school system was a natural focal point for bringing together the two segments of the population.

In order to coordinate a reading program based on diagnostic methods, and to provide new materials, Desert Basin implemented a "Reading" component in grades 1–8. For preschool-age children, "Early Childhood" classes in the elementary school provided reading-readiness and related skills taught by a full-time coordinator and a full-time aide.

At all schools, some "Bilingual" instruction was given to Spanish-speaking students, using many locally developed materials.

Coordination of after-school programs, recreation activities, summer school, and adult education was provided by the "Community Schools" component and its full-time coordinator. The "Counseling and Guidance" component served the elementary and middle schools and later covered the high school as well. Two full-time counselors were employed under this component.

The "Administrative" component was intended to make the project efficient in the use of its administrative resources, and involved the coordinating by the superintendent and the ES project director of their time and activities. "Evaluation" became the responsibility of a committee of five teachers and one community resident who met weekly to evaluate the project. Among their activities was the development of reading tests which were given to all students.

Most well known in the Desert Basin community was the "Media" component, which trained students in the use of media and actually reached into the community via the school's cable-television station. This component involved students in grades 7–12 as well as community groups.

In the spring of 1976, all components of Desert Basin's ES project continued to be in operation.

Desert Basin ranked "moderately low" on continuation. The district was particularly dissatisfied with one of its major components, "Bilingual Education," and in large part because of its desire to discontinue it, terminated its relationship with ES/Washington one year early. Many of its remaining components were largely reduced in scope, such as the "Media" program. However, it is claimed that more teachers are using individualized instruction than previously, and that there is a raised consciousness on the part of teachers toward the needs of students as individuals. The impacts of the project are felt most in the reading program in the elementary school, although the "Community Schools" program of adult education and recreation continues to be popular.

APPENDIX B

Instruments

PROFESSIONAL PERSONNEL CENSUS FORM

1. **What is the name of the school in which you are currently employed? (If you are currently employed in the central office and not in a particular school, please write "NONE." If you are currently employed in more than one school, please write the name of the school in which you spend the *most* time. If you spend an *equal* amount of time in more than one school, please write the name of each school.)**

   ____
   (name of school)

2. **During the current school year, approximately what percent of your professional time is spent as a . . .**

   a. Teacher.. ___ %
   b. Administrator .... ___ %
   c. Other (please specify): ______________________ ...... ___ %
   Total Professional Time 100 %

3. **Please circle those grade levels or ungraded levels that you *are currently teaching.* (If you are a teaching specialist or coordinator, please circle appropriate grade levels.) Then for each grade level which you are currently teaching, please write in the *number* of pupils you are teaching in each of these grades.**

Grade:	Pre-School	K	1	2	3	4	5	6	7	8	9	10	11	12
No. of Pupils	___	___	___	___	___	___	___	___	___	___	___	___	___	___

Ungraded Level:	Ungraded Elementary	Ungraded Middle or Jr. High	Ungraded High School
No. of Pupils	___	___	___

4. **For each subject area that you teach during the present school year, please answer the following questions:**
   1) Approximately how many hours per week do you teach the subject? (If none, write "0")
   2) Has the teaching of that subject area during the school year been affected by the Experimental Schools (ES) project in any of the following ways: curriculum content, instructional materials, teaching techniques, or in other ways? If "other," please specify. (To answer, please write the appropriate number in the boxes.)

Subject Area	Hours per week (if none, write "0")	Has ES affected the teaching of the subject in any of the following ways? (write appropriate numbers in boxes below) 3 = yes, greatly; 2 = yes, partly; 1 = no; ? = don't know — Curriculum Content	Instructional Materials	Teaching Techniques	Other (please specify):
a. Reading	____ hrs.	☐	☐	☐	☐ ______
b. Language Arts/ English	____ hrs.	☐	☐	☐	☐ ______
c. Social Studies	____ hrs.	☐	☐	☐	☐ ______
d. Mathematics	____ hrs	☐	☐	☐	☐ ______
e. Science	____ hrs.	☐	☐	☐	☐ ______
f. Foreign Languages	____ hrs.	☐	☐	☐	☐ ______
g. Industrial Arts	____ hrs.	☐	☐	☐	☐ ______
h. Agricultural Arts	____ hrs.	☐	☐	☐	☐ ______
i. Home Economics	____ hrs.	☐	☐	☐	☐ ______
j. Fine Arts	____ hrs.	☐	☐	☐	☐ ______
k. Business or Commercial	____ hrs.	☐	☐	☐	☐ ______
l. Physical Education	____ hrs.	☐	☐	☐	☐ ______
m. Remedial Reading	____ hrs.	☐	☐	☐	☐ ______
n. Athletic Coaching	____ hrs.	☐	☐	☐	☐ ______
o. Other (please specify): ______	hrs.	☐	☐	☐	☐ ______

5. A teacher's working day consists of many instructional and non-instructional activities. Not every teacher carries out the same activities, nor does every teacher spend the same amount of time on these activities. The Experimental Schools project in your school may affect these different activities and the way in which you spend your time.

For each activity listed below, please answer these questions:

1) How many minutes, if any, do you spend in a typical day on each activity? (if none, write "0.")
2) Is the activity affected by your school's participation in the Experimental Schools project?

Activity	Minutes per day *you* spend on this activity – (if none, write "0.")	Is this activity affected by your school's participation in the ES project? (circle one)
a. Planning class assignments and projects	______ min./day	greatly partly no ?
b. Locating and assembling instructional materials for class use	______ min./day	greatly partly no ?
c. Correcting assignments and written work	______ min./day	greatly partly no ?
d. Keeping records of pupil progress and grades	______ min./day	greatly partly no ?
e. Holding pupil conferences	______ min./day	greatly partly no ?
f. Handling discipline and behavior problems	______ min./day	greatly partly no ?
g. Administering teacher-made or standardized tests	______ min./day	greatly partly no ?
h. Guiding the class (as a whole) in discussions or assignments	______ min./day	greatly partly no ?
i. Making up tests and exercises for the class	______ min./day	greatly partly no ?
j. Working with individual pupils on their learning problems	______ min./day	greatly partly no ?
k. Instructing or working with small groups of pupils	______ min./day	greatly partly no ?
l. Being available while pupils work independently	______ min./day	greatly partly no ?
m. Doing diagnostic work on the learning difficulties of individual pupils	______ min./day	greatly partly no ?
n. Helping pupils plan their own special studies and projects	______ min./day	greatly partly no ?
o. Holding remedial sessions with pupils	______ min./day	greatly partly no ?
p. Coordinating activities with other teachers	______ min./day	greatly partly no ?
q. Doing bookkeeping chores (checking textbooks, attendance, lunch money)	______ min./day	greatly partly no ?
r. Typing or duplicating material to be used in class	______ min./day	greatly partly no ?
s. Conducting "housekeeping" chores (room cleaning and straightening, bulletin boards, etc.)	______ min./day	greatly partly no ?
t. Monitoring hallways, playground, lunchroom, etc.	______ min./day	greatly partly no ?

6. Are you supervising any *practice teachers* this term?

(Circle one)

Yes . . . . .1 → If yes, how many? ____ persons

No. 2

How many total hours per week are these practice teachers with you in the classroom? ____ hours

7. Do you have any teaching aides who work with you?

(Circle one)

No. . . . . .2 → *If no, skip to question 8.*

Yes 1

If yes, how many?. . . . . ____ persons

Of these persons, how many are paid employees?. . . . . ____ persons

How many are volunteers? . . . . ____ persons

How many hours per week do you have at least one teaching aide who works with you? . . . . ____ hours

How many teaching aides worked with you last year? (If none, write "0". If you were not teaching in this school last year, write "NA".). . . . . ____ persons

8. In your opinion, was it a good idea for this school district to participate in the Experimental Schools Program?

(Circle one)

It was a good idea 1

It was *not* a good idea 0

I have no opinion on this. . . . .?

9. In your opinion, what has been the overall effect of the ES project in your school?

(Circle one)

Very positive effect. 5

Somewhat positive effect 4

No effect. 3

Somewhat negative effect 2

Very negative effect 1

10. **The monies which are currently being supplied to your school district by the Experimental Schools Program in Washington, D.C. are scheduled to end in the near future. In your opinion, what effort should be made by your school district to continue the activities which are now being paid for by these federal funds?**

(Circle one)

Once the federal funds are gone, the activities should be continued, even if it means a slight increase in school taxes 3

Once the federal funds are gone, the activities should be continued, but only if it does not mean an increase in school taxes 2

Once the federal funds are gone, these activities should be discontinued 1

11. **In your opinion, would it be a good idea for this school district to participate in a federally funded project in the future?**

(Circle one)

It would be a good idea. 1

It would *not* be a good idea. 0

I have no opinion on this. . . . . ?

12. **In your judgment, which of the following statements *best reflects* the cumulative impact of your district's ES project upon the content of education in this school in the 1976-1977 school year?**

(Circle one)

ES funded activities resulted in *no change* in the curriculum. 0

ES funded activities resulted in a *shift in the emphasis* placed upon various elements of the curriculum or the addition of a few new curriculum elements 1

ES funded activities resulted in the *addition* of many new curricular elements 2

ES funded activities resulted in the *substitution* of many new curricular elements for previous curricular elements. 3

13. **In your judgment, which one of the following statements *best reflects* the cumulative impact of your district's ES project upon the structure and/or organization of this school?**

(Circle one)

ES funded activities resulted in *no change* in the organization of this school. 0

ES funded activites resulted in structural or organizational *changes primarily at the classroom level;* the organization of the school as a whole has not changed much. 1

ES funded activities have resulted in structural or organizational changes at the *school level* but there have been few changes at the classroom level 2

ES funded activities have resulted in structural or organizational *changes affecting the entire school* including the classroom level 3

14. Which of the following factors were features of the Experimental Schools project in your school district? For each factor checked "Yes" please indicate how the factor affected the implementation of the Experimental Schools project.

Factor	Was this a feature of the ES Project? No Yes (circle one)	What effect did this factor have on ES implementation? 5 = Strongly aided 4 = Moderately aided 3 = No effect 2 = Moderately impeded 1 = Strongly impeded (circle one)
a) the acquisition of new materials	2..... 1...	... 5....... 4 ...... 3 ...... 2 1
b) special training of new staff	2..... 1...	... 5....... 4...... 3 .... 2 1
c) the addition to our staff of people with special skills	2 .... 1...	... 5....... 4 ...... 3 .... 2 1
d) the use of outside consultants	2 .....1 ...	....5 ...... 4... .. 3 .. ....2 1
e) participation of teachers in ES planning	2 ....1 ...	... 5 ...... 4... .. 3 .. ....2 1

15. How have the following additional factors affected the implementation of the ES project in your school district?

Factor	Effect on ES Implementation 5 = Strongly aided 4 = Moderately aided 3 = No effect 2 = Moderately impeded 1 = Strongly impeded
a) the level of support of teachers for the project	5....... 4 ...... 3 ....2..........1
b) the level of support of the community for the project	5....... 4 ...... 3 ....2.... 1
c) the level of support of pupils for the project	5....... 4 ...... 3 ....2.... 1
d) conflict in the school district prior to ES	5....... 4 ...... 3 ....2.... 1
e) conflict in the school district caused by ES	5....... 4 ...... 3 ....2.... 1
f) the level of staff training prior to participation in ES	5....... 4 ...... 3 ....2.... 1
g) the level of expertise in the school district for the planning and implementation of a large project such as ES	5....... 4 ...... 3 ....2 1
h) the turnover of professional staff	5....... 4 ...... 3 ....2.......... 1
i) the paperwork involved in the project	5....... 4 ...... 3 ....2..........1
j) the quality of communication between the federal representatives and local personnel	5....... 4 ...... 3 ....2.... 1
k) the understanding of local situations on the part of the federal government	5....... 4....... 3 .... 2.... 1

16. A variety of instructional materials are used by teachers to accomplish different instructional objectives. A) During a typical school week, how many hours does the *typical* pupil in your class use each of the following types of materials? B) How much has the use of these instructional materials been affected by your school's participation in the Experimental Schools program?

Instructional Materials	Hours Per Week Materials Used	Effect of ES 5 = Greatly increased use 4 = Somewhat increased use 3 = No change 2 = Somewhat decreased use 1 = Greatly decreased use (Circle one)
a. Watching instructional television	____ hrs/wk	5 . . . . . .4 . . . . . . 3. . . . . . 2 1
b. Listening to tapes or records	____ hrs/wk	5 . . . . . .4 . . . . . . 3. . . . . . 2 1
c. Watching films or filmstrips	____ hrs/wk	5 . . . . . .4 . . . . . . 3. . . . . . 2 1
d. Working with a programmed textbook	____ hrs/wk	5 . . . . . .4 . . . . . . 3. . . . . . 2 1
e. Working at a computer terminal	____ hrs/wk	5 . . . . . .4 . . . . . . 3. . . . . . 2 1
f. Reading a conventional book or textbook	____ hrs/wk	5 . . . . . .4 . . . . . . 3. . . . . . 2 1
g. Writing on printed workbooks or worksheets prepared by a commercial publisher	____ hrs/wk	5 . . . . . .4 . . . . . . 3. . . . . . 2 1
h. Writing on workbooks or worksheets prepared by *this school system*	____ hrs/wk	5 . . . . . .4 . . . . . . 3. . . . . . 2 1
i. Writing on workbooks or worksheets prepared *by you or another teacher* in this school	____ hrs/wk	5 . . . . . .4 . . . . . . 3. . . . . . 2 1

17. Teachers vary in the variety of instructional materials which they use in their classes at any particular time. On some occasions they use the same materials with all their pupils, while on other occasions they assign different materials to different pupils according to their abilities, interests, etc. For each of the following types of instructional situations, please indicate 1) How frequently the pupils in your classes are using *different* materials and 2) How the frequency of these instructional situations has been affected by your district's Experimental Schools (ES) project. (Please write appropriate number in the boxes.)

Instructional Situations When pupils in your class use the following materials, how frequently are some pupils using . . .	Frequency of Use 5 = Always 4 = Almost always 3 = Frequently 2 = Occasionally 1 = Almost never 0 = Never	Effect of ES Project 5 = Greatly increased frequency 4 = Somewhat increased frequency 3 = No change 2 = Somewhat decreased frequency 1 = Greatly decreased frequency
a. Published *workbooks* that are different from what other pupils in your class are using	☐	☐
b. Teacher-prepared *worksheets* that are different from what other pupils in your class are using	☐	☐
c. Published *textbooks* that are different from what other pupils in your class are using	☐	☐
d. Teacher-prepared *tests* that are different from what other pupils in your class are using	☐	☐
e. Published *tapes* or *records* that are different from what other pupils are using	☐	☐

18. How often in your classes do you . . .

(Circle one on each line.)

Activity	Never	Almost Never	Occasionally	Almost Always	Always
a. Have all pupils work from the same text or workbook	1	2	3	4	5
b. Have pupils work in small groups	1	2	3	4	5
c. Work alone with individual pupils	1	2	3	4	5
d. Use audio-visual materials	1	2	3	4	5
e. Permit pupils to leave class when they have finished their work	1	2	3	4	5
f. Let pupils move freely about the room during class	1	2	3	4	5
g. Let pupils select where they will sit	1	2	3	4	5
h. Let pupils hand in an assignment late without penalty	1	2	3	4	5

19. During the current year, how many of the following activities did you participate in? How many of these activities were a part of your school district's Experimental Schools project?

Activity	How many of these did you participate in?	How many of these were part of your ES project?
a. Formal committees in your school	_____ committees	_____ committees
b. Formal committees in the school district but not limited to your school	_____ committees	_____ committees
c. In-service training or workshops during the summer of 1976	_____ workshops	_____ workshops
d. Professional trips	_____ trips	_____ trips
e. Field trips with pupils	_____ trips	_____ trips
f. Meetings with parent groups	_____ meetings	_____ meetings
g. Meetings with pupil groups	_____ meetings	_____ meetings
h. Meetings with community groups	_____ meetings	_____ meetings

20. There are many possible goals of public school education. The following is not a list of all possible goals, but would you please rate each goal in terms of a) its *importance* as you see it, b) how well your school is accomplishing it; and c) the effect the Experimental Schools (ES) program had in the accomplishment of this goal. (Please write appropriate number in the boxes.)

Goals	How *important* is this goal?	How well is your school accomplishing this goal?	What effects has the ES program had on the accomplishment of this goal?
	3 = Very important 2 = Moderately important 1 = Slightly important 0 = Not important ? = I don't know	3 = Very well 2 = Moderately well 1 = Slightly well 0 = Not well at all ? = I don't know	5 = very positive effect 4 = moderately positive effect 3 = no effect 2 = moderately negative effect 1 = very negative effect ? = I don't know
a. That students have such basic skills as reading, writing, and arithmetic	☐	☐	☐
b. That students respect authority	☐	☐	☐
c. That students have skills that will help them to get good jobs after they graduate from high school	☐	☐	☐
d. That students have the skills needed to succeed in college	☐	☐	☐
e. That students live moral lives	☐	☐	☐
f. That students be good citizens	☐	☐	☐
g. That students get along well with people in a variety of occupations	☐	☐	☐
h. That students get along well with people of all racial and ethnic backgrounds	☐	☐	☐
i. That students develop an understanding and appreciation of the philosophy and thought of man	☐	☐	☐
j. That students think for themselves	☐	☐	☐
k. That students get an education geared to their individual needs	☐	☐	☐
l. That students can be what they choose to be in terms of career and lifestyle	☐	☐	☐

21. Of the goals listed above (a-l), which one do you consider to be:

a. *Most important:* a b c d e f g h i j k l (Circle one)

b. *Best accomplished by this school:* a b c d e f g h i j k l (Circle one)

c. *Goal most positively affected by ES program:* a b c d e f g h i j k l (Circle one)

d. *Goal most negatively affected by ES program:* a b c d e f g h i j k l (Circle one)

22. In most schools there are a variety of problems which must be solved if the school is to do a better job. Listed below is a series of such problems. a) To what extent is each of these potential problems actually a problem in your school? b) To what degree has your ES project helped solve this problem? (Please write appropriate number in the boxes.)

Potential Problem	3 = This is a serious problem in my school 2 = This is a moderate problem in my school 1 = This is a minor problem in my school 0 = This is not a problem in my school ? = I don't know	To what degree has the ES project helped solve this problem? 5 = greatly helped 4 = moderately helped 3 = no effect 2 = moderately impeded 1 = greatly impeded
a. Our physical plant is inadequate	☐	☐
b. We do not have enough appropriate instructional materials and supplies	☐	☐
c. We have too many teachers who do not have the teaching skill needed by our pupils	☐	☐
d. We have too many teachers who do not have the motivation necessary to teach our pupils	☐	☐
e. We have too many pupils who do not have the academic ability necessary to succeed in school	☐	☐
f. We have too many pupils who do not have the motivation necessary to succeed in school	☐	☐
g. We do not have enough information about instructional materials and techniques which might be better than the ones which we are currently using	☐	☐
h. We do not have enough information about how well we are achieving our educational goals	☐	☐
i. Too few of our graduates are able to succeed in whatever school or college they go to next	☐	☐
j. Too few of our graduates are able to succeed in whatever employment they undertake	☐	☐
k. Our school is too small to offer the type of educational program needed by our pupils	☐	☐
l. Our school is not organized in a way which uses the available resources most effectively	☐	☐
m. Our school places too much emphasis upon the way things have been done in the past	☐	☐
n. We don't know enough about our community's true concerns and preferences regarding what our schools should be doing	☐	☐
o. We are not responsive enough to what the community tells us	☐	☐

23. Listed below are a series of statements about which professional personnel in rural American school districts are likely to hold differing opinions. Please indicate the extent to which you agree or disagree with these statements by *circling* the appropriate abbreviation. (There are no "right" or "wrong" answers to these statements. Please give your *personal opinion.*)

Statement	I Strongly Disagree	I Disagree	I am Undecided	I Agree	I Strongly Agree
a. There is too much change in today's world.	SD	D	U	A	SA
b. There is really no need for this school system to be doing things any differently now than it did in the past	SD	D	U	A	SA
c. Pupil dress codes should be strictly enforced.	SD	D	U	A	SA
d. When a teacher is friendly with pupils they are likely to misbehave	SD	D	U	A	SA
e. Only teachers who want to change their teaching methods should be asked to do so.	SD	D	U	A	SA
f. The best teaching methods are those which have stood the test of time	SD	D	U	A	SA
g. When pupils are allowed to arrange their desks according to their own preferences, they usually abuse the privilege	SD	D	U	A	SA
h. When you try to change the way things are done in rural schools, you usually make them worse	SD	D	U	A	SA
i. A school system should not try out any new ideas unless there is definite proof that they will work better than present practices	SD	D	U	A	SA
j. A teacher should always be present when pupils are taking examinations	SD	D	U	A	SA
k. Pupils should be assigned to specific seats in their classrooms	SD	D	U	A	SA
l. Trying to change a school's curriculum is generally more trouble than it is worth	SD	D	U	A	SA
m. There are a lot of new ideas being tried out in other school systems which we ought to be trying in this school	SD	D	U	A	SA
n. Effective learning seldom takes place in a noisy classroom	SD	D	U	A	SA
o. It is more important for pupils to learn to obey rules than to make their own decisions.	SD	D	U	A	SA
p. When pupils are allowed to use the lavatory during class periods they usually abuse the privilege	SD	D	U	A	SA
q. New ideas ought to be tried in a single classroom before they are introduced throughout an entire school system	SD	D	U	A	SA

23. (continued)

Statement	I Strongly Disagree	I Disagree	I am Undecided	I Agree	I Strongly Agree
r. Teaching is most effective when all pupils are working on the same subject matter	SD	D	U	A	SA
s. The present curriculum of this school system makes adequate provision for difference among pupils	SD	D	U	A	SA
t. Children should be allowed to exercise considerable choice in what they learn	SD	D	U	A	SA
u. A teacher's style should vary with the needs of pupils	SD	D	U	A	SA
v. A pupil who deliberately destroys school property should be severely punished	SD	D	U	A	SA
w. When we make improvements in our schools, the community never knows it	SD	D	U	A	SA
x. It is hard to have strong common support for schools without continuous parent and community group involvement	SD	D	U	A	SA

24. During a typical school year, many decisions must be made. Not all people influence any particular decision, and the degree of influence of different persons generally varies with the practices being decided upon. Please indicate, *in your opinion,* the degree of influence each of the persons listed below has on the following decisions.

0 = Usually has *no* influence
1 = Usually has *minor* influence
2 = Usually has *moderate* influence
3 = Usually has *decisive* influence
? = I don't know

(Please insert the appropriate code number in the appropriate box.)

Decisions	Persons			
	School Board Members	Superintendent	Principals	Teachers
a. Selecting required texts and other materials	☐	☐	☐	☐
b. Establishing the objectives for each course	☐	☐	☐	☐
c. Determining the concepts and information to be taught in a particular day	☐	☐	☐	☐
d. Determining daily lesson plans and activities	☐	☐	☐	☐
e. Adding or dropping courses	☐	☐	☐	☐
f. Hiring of new teachers	☐	☐	☐	☐
g. Deciding whether to renew a teacher's contract	☐	☐	☐	☐
h. Making specific faculty assignments	☐	☐	☐	☐
i. Planning new buildings and facilities	☐	☐	☐	☐
j. Establishing salary schedules	☐	☐	☐	☐
k. Setting the time schedule and goals to be achieved for a system-wide change	☐	☐	☐	☐
l. Identifying types of system-wide changes to be implemented	☐	☐	☐	☐
m. Working out details for implementing system-wide changes	☐	☐	☐	☐

25. Now please think in terms of your *own* degree of influence in these same decisions and indicate 1) how much influence you have in these decisions, and 2) how much influence you feel you *should* have. (In answering these questions, please circle the appropriate number.)

Decisions	How much influence *do* you have?					How much influence *should* you have?				
	0 = No influence 1 = Very little influence 2 = Some influence 3 = Moderate influence 4 = A great deal of influence									
	(Circle one)					(Circle one)				
a. Selecting required texts and other materials	0	1	2	3	4	0	1	2	3	4
b. Establishing the objectives for each course	0	1	2	3	4	0	1	2	3	4
c. Determining the concepts and information to be taught in a particular course	0	1	2	3	4	0	1	2	3	4
d. Determining daily lesson plans and activities	0	1	2	3	4	0	1	2	3	4
e. Adding or dropping courses	0	1	2	3	4	0	1	2	3	4
f. Hiring new teachers	0	1	2	3	4	0	1	2	3	4
g. Deciding whether to renew a teacher's contract	0	1	2	3	4	0	1	2	3	4
h. Making specific faculty assignments	0	1	2	3	4	0	1	2	3	4
i. Planning new buildings and facilities	0	1	2	3	4	0	1	2	3	4
j. Establishing salary schedules	0	1	2	3	4	0	1	2	3	4
k. Setting the time schedule and goals to be achieved for a system-wide change	0	1	2	3	4	0	1	2	3	4
l. Identifying types of system-wide changes to be implemented	0	1	2	3	4	0	1	2	3	4
m. Working out details for implementing system-wide changes	0	1	2	3	4	0	1	2	3	4

26. a) In your opinion, how much influence did each of the following persons or groups have in *implementing* the Experimental Schools project in your school district?

b) How much influence *should* each have had?

c) How much has the influence of this person or group in the decision-making process in your school changed because of ES?

Type of Person or Group	How much influence *did* each have? 4 = A great deal 3 = Some 2 = Very little 1 = Almost none 0 = None ? = I don't know	How much influence *should* each have had? 4 = A great deal 3 = Some 2 = Very little 1 = Almost none 0 = None ? = I don't know	How much did the influence of each change because of ES? 4 = Big increase 3 = Some increase 2 = None or little 1 = Some decrease 0 = Big decrease ? = I don't know
a. The school board	☐	☐	☐
b. The superintendent	☐	☐	☐
c. Central office staff other than the superintendent	☐	☐	☐
d. Principals	☐	☐	☐
e. Teachers	☐	☐	☐
f. Teacher aides	☐	☐	☐
g. Parents	☐	☐	☐
h. Students	☐	☐	☐
i. Community groups	☐	☐	☐
j. Outside consultants	☐	☐	☐
k. Staff of the Experimental Schools Program in Washington, D.C.	☐	☐	☐
l. Others (please specify) ______ ______	☐	☐	☐

27. Listed below are a series of policy areas in which policies may or may not exist *in your school.* For each policy statement, please indicate: 1) Whether such a policy exists, 2) How often this policy is enforced? (Please write the appropriate number in each box.)

Policy Area	Is there a policy in this area? 2 = Yes, a written one 1 = Yes, an unwritten one 0 = No such policy exists ? = I don't know	If a policy exists, how often is this policy enforced? 5 = Always 4 = Almost always 3 = Frequently 2 = Occasionally 1 = Almost Never 0 = Never ? = I don't know
a. Pupil appearance	☐	☐
b. Pupil smoking	☐	☐
c. Pupil drinking	☐	☐
d. Pupil romancing	☐	☐
e. Grading of pupils	☐	☐
f. Disciplining of pupils	☐	☐
g. Teacher use of curriculum guides	☐	☐
h. Teacher preparation of lesson plans	☐	☐
i. Teacher salary increases	☐	☐
j. Teacher tenure	☐	☐
k. Arrival and departure time for teachers	☐	☐
l. Teacher responsibilities for extra-curricular activities	☐	☐
m. Teacher responsibilities for parent-teacher activities	☐	☐
n. Field trips	☐	☐
o. Textbook selection	☐	☐
p. Invitation of outside speakers into classes	☐	☐
q. Hiring of new teachers	☐	☐
r. Dismissal of present teachers	☐	☐

28. Listed below are a series of statements about activities which may or may not occur in your school. Please read each event statement carefully and then indicate 1) whether that activity is characteristic of your school and 2) whether this event has been affected by your schools' participation in the Experimental Schools program.

Activity	Activity is Characteristic of the School 3 = Yes 2 = Generally yes 1 = Generally no 0 = No	Effect of ES Program 3 = Occurs more because of ES 2 = Not affected by ES 1 = Occurs less because of ES
a. Teachers may arrange the desks in their classrooms however they like	☐	☐
b. Teachers may plan their instructional periods however they like	☐	☐
c. Teachers may make their own decisions about problems that come up in their classrooms	☐	☐
d. Teachers may try out new ideas in their classes	☐	☐
e. Teachers may set their own standards for what grades are given for various levels of pupil performance	☐	☐
f. Teachers can plan and schedule their own field trips	☐	☐
g. Teachers can make their own selection of textbooks	☐	☐
h. Teachers may decide on the format for their lesson plans	☐	☐
i. Teachers may determine how best to achieve their course objectives	☐	☐
j. Teachers may exercise the final authority over major instructional decisions	☐	☐
k. Teachers feel free to call on other teachers for help in solving their problems	☐	☐
l. There is a general consensus among teachers about the way the school should be run	☐	☐
m. Experienced teachers accept new or younger teachers as colleagues	☐	☐
n. Teachers cooperate with each other to achieve common personal and professional goals	☐	☐
o. Teachers of various classes try to avoid creating problems for each other	☐	☐
p. Teachers feel free to call on administrators for help in solving problems	☐	☐
q. The people in this community respect their teachers and treat them as professionals	☐	☐

28. (continued)

Activity	Activity is Characteristic of the School 3 = Yes 2 = Generally yes 1 = Generally no 0 = No	Effect of ES Program 3 = Occurs more because of ES 2 = Not affected by ES 1 = Occurs less because of ES
r. The people in this community appreciate the schools and what they are doing	☐	☐
s. The people in this community make the teachers feel as if they are a real part of the community	☐	☐
t. Special orientation meetings are held for new teachers	☐	☐
u. Special social gatherings are held to welcome new teachers	☐	☐
v. New teachers are given a handbook to acquaint them with the school	☐	☐
w. Experienced teachers are assigned to new teachers as "buddies" to help acquaint them with the school and its procedures	☐	☐
x. Teachers are encouraged to meet on a regular basis with parents	☐	☐
y. Teachers are encouraged to participate in community groups in order that they understand what the community wants from their schools	☐	☐
z. Teachers often work together on team teaching efforts in the classroom	☐	☐
aa. Teachers often participate in joint planning for educational activities	☐	☐

29. In most schools, specific issues or events may occur over which there are differences of opinion resulting in *disputes*. During the last 12 months, how frequently have disputes among teachers and between teachers and others occurred at your school, regarding the following issues and event?

4 = Very frequently
3 = Frequently
2 = Occasionally
1 = Rarely
0 = Never
? = I don't know

(In answering these questions, please insert the appropriate number in each box)

Issues and Events	How frequently have disputes occurred . . .		
	Between Teachers	Between teachers & administrators	Between teachers & students
a. The initial hiring of a teacher	☐	☐	☐
b. The dismissal of a teacher	☐	☐	☐
c. The introduction of a curriculum change	☐	☐	☐
d. The retention of a textbook	☐	☐	☐
e. The teaching of controversial material	☐	☐	☐
f. Student dress or appearance	☐	☐	☐
g. Student moral behavior	☐	☐	☐
h. Teacher dress or appearance	☐	☐	☐
i. Teacher moral behavior	☐	☐	☐
j. The dismissal of students from class for extracurricular activities	☐	☐	☐
k. Salary decisions	☐	☐	☐
l. Teacher course assignments	☐	☐	☐
m. Teacher participation in non-teaching duties (e.g., lunchroom duty, bus duty, etc.)	☐	☐	☐
n. Teacher participation in extracurricular activities (e.g., clubs, dances, athletic events)	☐	☐	☐
o. The use of teachers as substitutes for other absent teachers	☐	☐	☐
p. Teacher participation in evening or weekend meetings or activities	☐	☐	☐
q. The need for administrative support in handling pupil behavior problems	☐	☐	☐
r. The need for administrative support in dealing with parents	☐	☐	☐
s. Excessive "red tape" when trying to get a decision	☐	☐	☐
t. The need to do unnecessary "paper work"	☐	☐	☐
u. Determination of appropriate homework assignments	☐	☐	☐
v. The necessity for rules and regulations	☐	☐	☐
w. The schools responsibilities to the community	☐	☐	☐

30. In all communities there is, from time to time, tension between various groups who have different ideas about how things should be done. In general, how much tension *regarding matters at your school* do you feel currently exists between each of the following *groups* of people? (In answering this question, please think of teachers, students, parents, etc., as groups, even though not *all* members of a particular group necessarily feel the same way.)

How much tension currently exists between . . .	Degree of tension				
	Much Tension	Some Tension	Very Little Tension	No Tension	I don't Know
	(Circle one on *each* line)				
a. Different groups of teachers?	3	2	1	0	?
b. Teachers and the principal?	3	2	1	0	?
c. Teachers and the superintendent?	3	2	1	0	?
d. Teachers and the central office staff?	3	2	1	0	?
e. Teachers and the school board?	3	2	1	0	?
f. Teachers and students?	3	2	1	0	?
g. Teachers and parents?	3	2	1	0	?
h. Teachers and community groups?	3	2	1	0	?
i. The principal and the superintendent?	3	2	1	0	?
j. The principal and the central office staff?	3	2	1	0	?
k. The principal and the school board?	3	2	1	0	?
l. The principal and the students?	3	2	1	0	?
m. The principal and parents?	3	2	1	0	?
n. The principal and community groups	3	2	1	0	?
o. The superintendent and central office staff?	3	2	1	0	?
p. The superintendent and the school board?	3	2	1	0	?
q. The superintendent and students?	3	2	1	0	?
r. The superintendent and community groups?	3	2	1	0	?
s. The school board and students?	3	2	1	0	?
t. The school board and parents?	3	2	1	0	?
u. The school board and community groups?	3	2	1	0	?

31. In which of the following professional organizations do you currently hold membership? (Please check as many as apply.)

Professional Organizations	Check organization to which you belong	For each organization checked: How many meetings have you attended in the past 12 months?	For each organization checked: Have you ever held office in this organization 2 = Yes; 1 = No
a. Our *local* classroom teachers association affiliated with NEA	____	☐	☐
b. Our *state* classroom teachers association affiliated with NEA	____	☐	☐
c. The National Education Association (NEA)	____	☐	☐
d. Our local classroom teachers association affiliated with AFT	____	☐	☐
e. Our *state* classroom teachers association affiliated with AFT	____	☐	☐
f. The American Federation of Teachers (AFT)	____	☐	☐
g. The national teacher organization in my subject area	____	☐	☐
h. A national education honorary society	____	☐	☐
i. A national organization of school administrators	____	☐	☐
j. Our state organization of school administrators	____	☐	☐
k. Other (Please specify): ________________ ________________________	____	☐	☐
l. None	____		

32. What is the highest level of education which you have completed? (Circle one)

Less than a bachelor's degree 1

Bachelor's degree 2

Bachelor's degree plus 15 or more hours 3

Master's degree 4

Master's degree plus 30 or more hours 5

Doctorate 6

Other (specify) 7

33. How many years (including this year) have you had teaching or administrative experience *in this school district* (full-time or part-time)?

Number of years of teaching experience: ________ years

Number of years of administrative experience: ________ years

34. What is the *total* number of years (including this year) that you have been a *teacher* (full-time or part-time)? Include counseling and special education. How many of these years have been in school districts which are *larger* than the one in which you are currently employed and how many years have been in school districts that are smaller?

Total number of years ________

Years of experience in larger districts ________

Number of School Districts larger than yours in which you have had experience ________

Years of experience in smaller districts ________

Number of School Districts smaller than yours in which you have had experience ________

35. What is the *total* number of years (including this year) that you have had *administrative* experience (full-time or part-time)? How many of these years have been in school districts that are larger than the one in which you are currently employed and how many years have been in school districts that are smaller?

Total number of years ________

Years of experience in larger districts ________

Number of School Districts larger than yours in which you have had experience ________

Years of experience in smaller districts ________

Number of School Districts smaller than yours in which you have had experience ________

36. What is your sex? (Circle one)

Male 1

Female 2

37. When were you born?

Year ________

Thank you for your assistance. Please mail this completed questionnaire in the enclosed postpaid envelope by May 1, 1977.

ON-SITE RESEARCHER QUESTIONNAIRE

Part II: School Form

Scope of Implementation of ES Projects (1975-1976)

1. Name of School: ______________________________
   (a) Grades included in this school: (Circle all grades that apply)
   K 1 2 3 4 5 6 7 8 9 10 12 Ungraded
   (b) Number of full-time teachers: _____
   (c) Number of part-time teachers: _____
   (d) Number of teacher aides (full-time and part-time): _____
   (e) Number of administrators: _____
   (f) Number of school-based specialists (full-time and part-time): _____
   (g) Number of pupils: _____
   (h) Length of school day (to nearest half hour): _____
   (i) Number of periods in school day (where applicable): _____
   (j) Who is the person responsible for coordinating ES in this school?
   ____
   (k) What is this person's title and/or position? (Check one)
   _____ Superintendent
   _____ Principal
   _____ Project Coordinator
   _____ Component Coordinator
   _____ Teacher
   _____ Specialist
   _____ Other
   (l) Is this person's primary base in this school? _____ Yes _____ No
   If No: Is this person based primarily:
   _____ In the district office
   _____ In several schools (splitting time among them)
   _____ Other (explain) ______________________
   (m) What percentage of this person's work week has been devoted in the past three years to supervising the ES project in this school?
   _____ % 1975-1976 _____ % 1974-1975 _____ % 1973-1974
2. Approximately how many students in this school are involved in ES funded activities
   _____ Students 1975-1976 _____ Students 1974-1975

3a. Thinking only of students who actually participate in ES funded activities in the 1975-1976 school year, please give us information on how much of a "typical" day these students spend in these activities by indicating how many of the students fall into each of the five categories defined below:

	Approximate number of students
81-100% of the typical day spent in ES funded activities	_____
61-80% of the typical day spent in ES funded activities	_____
41-60% of the typical day spent in ES funded activities	_____
21-40% of the typical day spent in ES funded activities	_____
1-20% of the typical day spent in ES funded activities	_____
TOTAL (all students involved in ES)	_____

3b. Does this degree of participation represent an increase, decrease or no change from the previous year (1974-75)? (Circle one number)

Large decrease	Small decrease	Same as last year	Small increase	Large increase
−2	−1	0	+1	+2

4. Approximately how many teachers in this school are involved in ES funded activities?

_____ Teachers 1975-1976 _____ Teachers 1974-1975

5a. Taking as a base those teachers who *do* participate in ES funded activities, during the 1975-1976 school year please indicate what percentage of these fall into each of the categories below:

	Approximate number of teachers
81-100% of the typical day spent in ES funded activities	_____
61-80% of the typical day spent in ES funded activities	_____
41-60% of the typical day spent in ES funded activities	_____
21-40% of the typical day spent in ES funded activities	_____
1-20% of the typical day spent in ES funded activities	_____
TOTAL (all teachers involved in ES)	_____

5b. Does this degree of participation represent an increase, decrease or no change from the previous year (1974-1975)? (Circle one number)

Large decrease	Small decrease	Same as last year	Small increase	Large increase
−2	−1	0	+1	+2
−2	−1	0	+1	+2

6. Please indicate the approximate percentage of the total time allocated to instruction at each grade level that was involved in ES funded activities

during the 1975-1976 school year. If the grade level was *not* involved at all, indicate "0%"; if the grade level does not exist in this school, indicate "NA"

K _____ %	5 _____ %	10 _____ %
1 _____ %	6 _____ %	11 _____ %
2 _____ %	7 _____ %	12 _____ %
3 _____ %	8 _____ %	
4 _____ %	9 _____ %	

7a. Please indicate the approximate percentage of the total time devoted to instruction in each subject area that was involved in ES funded activities during the 1975-1976 school year.
- If the subject area has been added as a result of ES, indicate 100% of the time is affected.
- If the subject area has *not* been involved at all, indicate "0%".
- If the subject area *does not exist* in this school, indicate "NA".

Subject area	Aproximate percentage of time affected
Reading	_____ %
Language Arts	_____ %
Social Studies	_____ %
Mathematics	_____ %
Sciences	_____ %
Languages (foreign)	_____ %
Industrial or Agricultural Arts	_____ %
Home Economics	_____ %
Fine Arts	_____ %
Business or Commercial	_____ %
Physical Education	_____ %
Other (Please Specify):	_____ %
______________________________	_____ %

7b. Does the total number of subject areas affected represent an increase, decrease, or no change from the previous year (1974-1975)?
(Circle one number)

Large decrease	Small decrease	Same as last year	Small increase	Large increase
−2	−1	0	+1	+2

7c. Does the percentage of time affected represent an increase, decrease or no change from the previous year (1974-1975)?
(Circle one number)

Large decrease	Small decrease	Same as last year	Small increase	Large increase
−2	−1	0	+1	+2

8a. What percentage of the classrooms or curricular units in this school have ES funded activities within them?
____ %
(If none, skip to question 9)

8b. For the ES affected classrooms noted above, which of the following statements most accurately reflects the degree to which the classrooms or curricular units are affected by the implementation of ES funded activities?
All are affected in a major way ____
Some are affected in a major way, others are affected in a moderate or minor way ____
All are affected in a moderate way ____
Some are affected in a moderate way, others are affected in a minor way ____
All are affected in only a minor way ____
Other ____

9. Relative to the amount of time, space and facilities devoted to formal education by this school prior to the 1972-1973 school year, to what degree have ES funded activities changed the use of time, space and facilities? (Circle one number on each line)

(a) Time

None	Very little				Very great
0	1	2	3	4	5

(b) Space

None	Very little				Very great
0	1	2	3	4	5

(c) Facilities

None	Very little				Very great
0	1	2	3	4	5

10. What effect have ES funded activities had on the degree of influence the following types of people have in the decision-making process within the school? (Check one on each line)

Types of people	Decreased influence		No change	Increased influence	
	Significantly	To some extent		To some extent	Significantly
Students					
Parents					
Teachers					
Principals					
Superintendent					
School Board					
Other Residents					

11. To what degree have ES funded activities in this *school* involved significant changes in the degree of involvement of community members as active participants in the educational process (e.g., as formal instructors within or outside the school classroom on a regular basis?) (Circle one number)

None	Very little				Very great
0	1	2	3	4	5

12. To what degree has community involvement in school meetings, PTA, Parents' nights, etc. been increased during the ES funding period? (Circle one number)

None	Very little				Very great
0	1	2	3	4	5

In your opinion, is this increase a result of ES stimulated efforts to encourage community participation?

_____ Yes _____ No _____ No increase

13. To what degree have ES funded activities in this school involved exchanging resources or developing joint programs with other schools or school districts? (Circle one number)

None	Very little				Very great
0	1	2	3	4	5

Has any effort in this area involved the movement of students from one school to another?

_____ Yes _____ No _____ No effort in this area

Has any effort in this area involved the exchange of staff resources?

_____ Yes _____ No _____ No effort in this area

14. To what degree have ES funded activities in this school produced changes in the following areas since the 1972-1973 school year? (Circle one number on each line)

(a) Change in the lines of authority

None	Very little				Very great
0	1	2	3	4	5

Describe ____

____

(b) Change in the decision-making power

None	Very little				Very great
0	1	2	3	4	5

Describe ____

____

(c) Changes in administrative procedure such as:

— Accounting procedures

None	Very little				Very great
0	1	2	3	4	5

Describe ____

____

— Record keeping

None	Very little				Very great
0	1	2	3	4	5

Describe____

____

— Other administrative procedures

None	Very little				Very great
0	1	2	3	4	5

Describe____

____

15. Please rank order the priority of the following facets of comprehensiveness in terms of the degree to which, in your opinion, they are emphasized by the administration and teachers of this school. Use a 1 to indicate the facet that is considered to be of greatest importance, and a 6 to indicate the facet that is given the least attention. Each rank should be used only once. (We are not concerned with the intent of the program planners and administrators, but with the actual emphasis, in terms of allocation of resources, at this point.)

Facets	Rank order of Administrators	Rank order of Teachers
Curriculum	_____	_____
Instruction and staffing	_____	_____
Use of time, space and facilities	_____	_____
School organization, administration and governance	_____	_____
Community participation	_____	_____
Ongoing evaluation	_____	_____

16. In your judgement, which of the following statement *best reflects* the impact of ES funded activities upon the content of education in this school in the 1975-1976 school year?

(Circle one)

(a) ES funded activities resulted in *no change* in the curriculum 0

(b) ES funded activities resulted in a *shift in the emphasis* placed upon various elements of the curriculum 1

(c) ES funded activities resulted in the *addition* of new curricular elements 2

(d) ES funded activities resulted in the *substitution* of new curricular elements, such that many older curricula have been replaced 3

17. In your judgment, which one of the following statements *best reflects* the impact of ES funded activities upon the structure and/or organization of this school?

(Circle one)

(a) ES funded activities resulted in *no change* in the organization of this school 0

(b) ES funded activities resulted in structural or organizational *changes primarily at the classroom level;* the organization of the school as a whole has not changed much 1

(c) ES funded activities have resulted in structural or organizational *changes affecting the entire school* that have replaced the previous structure of organization 2

18. For each ES component whose implementation has ever been attempted in this school, please indicate the status of the component for each of the school years below and how likely it is to persist after the period of ES funding is over.

Component	Year (For each year, indicate: N = Never attempted B = Begun C = Continued D = Discontinued			How likely are the activities associated with this component to be continued after the ES funding is over? 5 = Very likely 4 = Likely 3 = Unlikely 2 = Very unlikely 1 = Impossible to tell 0 = Already discontinued
	1973-74	1974-75	1975-76	

19. We are interested in finding out whether there are any other *new* programs that have been initiated in this school since 1973. If such programs exist, please describe them below, and indicate whether, in your judgment, they were stimulated by ES.

Were these programs:
Direct spinoffs from ES-funded activities = 4
Indirect spinoffs from ES-funded activities = 3
Unrelated to ES-funded activities = 2
Impossible to tell = 1
(Write appropriate number on each line in Column 2)

New Programs	Stimulated by ES

ON-SITE RESEARCHER QUESTIONNAIRE

Scope of Implementation of ES Components

Name of Component: ____________________

1. Does this component involve curriculum?

   _____ No (Continue with Question 2)

   _____ Yes (Continue with *a* to *d*)

   (a) Name all schools in which, or for which, this component had been implemented by May 1975. For each school, indicate the grade levels, approximate number of pupils involved, approximate number of teachers and teaching aides involved, and the approximate number of hours per week involved in the classroom.

Name of school	Grade level(s)	Number of pupils	Number of teachers	Number of teaching aides	Number of classroom hrs/week
1.					
2.					
3.					
4.					
5.					

   (b) For each of the schools listed above, are there any relevant class sections of the grade level(s) which you listed for which the component has *not* been implemented (Check one)

   No _____

   Yes _____ Indicate those grade levels in the previous question by circling them.

   (c) Approximately what percent of all classes in the school district were being affected by this component in May 1975? _____ %

   (d) Approximately what percent of all pupils in the school district were being affected by this component in May 1975? _____ %

2. Does this component involve staff development or training?
_____ No (Continue with Question 3)
_____ Yes (Continue with *a* to *c*)
(a) If the activities of this component vary across schools, please complete the following table. If it is an activity that is district-wide, please so indicate in the first column.

Name of school	Number of teachers	Number of teaching aides	Number of other individuals	Grade level(s) represented	Average number of hour each day/number of days
1.					
2.					
3.					
4.					
5.					

(b) Approximately what percent of teachers in the school district were affected by this component? _____ %
(c) Approximately what percent of the teachers for whom this component is relevant or applicable participated in it? (i.e., if it is a reading program and there are 10 teachers who teach reading and could be participating, what percentage of those 10 are actually involved?) _____ %

3. How much of the change intended by this component can be considered to be of the following types? (Circle one number on each line)
(a) New forms of *classroom organization,* including changes in teacher/pupil relations or teacher/teacher relations (for example, team teaching, individualized instruction).

None	Very little				Very great
0	1	2	3	4	5

(b) New *interpersonal skills* for staff (for example, sensitivity training).

None	Very little				Very great
0	1	2	3	4	5

(c) New *curricular ideas* or information.

None	Very little				Very great
0	1	2	3	4	5

4. Does this component involve community participation?
_____ No (continue with Question 5)
_____ Yes (continue with *a* & *b*)
(a) Please provide information for each of the categories of people involved in this component.

Category of people	Number of participants	Hours spent by typical participant during the year*
School professional personnel		
School nonprofessional personnel		
School board members		
Parents of school pupils		
Preschool children		
Senior citizens		
Other residents		

*If this is an ongoing activity, indicate the average number of hours per month (e.g., 5/mo. for "five hours per month")

(b) Would you characterize this form of "community involvement" as *primarily*...

(Circle one)

(1) Active involvement of community members (other than elective school officials) in the formal decision-making process of the district through committee assignments, task forces, etc? 6

(2) Active involvement of community members (other than certified teachers) in formal instructional process taking place within school classroom (e.g., guest teachers, mini-course teachers)? 5

(3) Active involvement of community members (other than certified teachers) in formal instructional processes taking place outside school classrooms (e.g., on field trips, as supervisors of on-the-job trainers, etc.) 4

(4) Active solicitation of ideas from community members through open meetings 3

(5) Active involvement of community members as students in continuing education classes 2

(6) Communicating school-related information to community members through speeches at local clubs, newsletters, newspaper articles, open meetings, etc. . 1

5. Does this component involve innovative uses of time, space or facilities?

_____ No (Continue with Question 6)

_____ Yes (Continue with *a* to *d*)

(a) Time: Relative to the amount of time devoted to formal education by this school district in the past, how much has this component involved change in use of *time*?

None	Very little				Very great
0	1	2	3	4	5

(b) Space: Relative to the amount of space devoted to formal education by this school district in the past, how much has this component involved change in use of *space*?

None	Very little				Very great
0	1	2	3	4	5

(c) Facilities: Relative to the amount of facilities devoted to formal education by this school district in the past, how much has this component involved change in the use of *facilities*?

None	Very little				Very great
0	1	2	3	4	5

(d) Has this component involved

(1) The purchase of new materials?

_____ No

_____ Yes. In general, have these materials been used?

_____ No

_____ Yes. Please describe new materials.

(2) A major change in the *use* of existing materials?

_____ No

_____ Yes. Please explain_____

6. On the chart below, please list by *position* all role incumbents who have a major responsibility in the actual implementation of this component. Then, also using the chart, answer questions *a* through *e* separately for each position. (Reminder: *a* through *e* are to be answered on the *chart*. Use the appropriate code numbers in answering each question.)

(a) How was the current incumbent in this position recruited?
Within the school district, but from another position = 1
The person has another position, but added these responsibilities to it = 2
From outside the district = 3

(b) Is this a position that has been created because of ES?
Yes = 1 (If 1, skip to Question *d*)
It was a previously existing position, but has been altered greatly by ES = 2
It was a previously existing position = 3

(c) To what extent has this person's responsibilities, or how this person carries them out, changed as a result of this component?
To a very great extent = 3; to some extent = 2; very little, if at all = 1

(d) What percent of this person's time is being devoted to this component _____ %

(e) Do the individual's responsibilities in this component require him or her to interact with

(1) Number of Individuals
More individuals than before? = 1
Fewer individuals than before? = 2

No difference in number of individuals = 3

(2) School People?
More types of people in the school system? = 1
Fewer types of people in the school system? = 2
No difference in the types of people in the school system = 3

(3) Community People?
More types of people from the local community? = 1
Fewer types of people from the local community? = 2
No difference in the types of people from the local community = 3

Position	(a) Recruit (1,2, or 3)	(b) Is position new (1,2, or 3)	(c) Responsibility change (1,2 or 3)	(d) Enter % time	(e) Change in interaction		
					No. individuals (1,2,3)	No. types School (1,2,3)	No. types in Community (1,2,3)

7. Have the following changes been associated with this component?
   (a) Change in the lines of authority
   _____ No
   _____ Yes. Describe ____________________
   (b) Change in decision-making power
   _____ No
   _____ Yes. Describe ____________________
   (c) Changes in administrative procedures such as:
   Accounting procedures _____ No
   _____ Yes. Describe ________________
   Record keeping _____ No
   _____ Yes. Describe ________________
   Other administrative procedures _____ No
   _____ Yes. Describe ________________

8. Has this component had any affect on the *influence* of the following types of people, on the decision making responsibility within the school district? (Check appropriate column for each row)

Types of people	Decreased influence		No change	Increased influence	
	Significantly	To some extent		Significantly	To some extent
Students					
Parents					
Teachers					
Principals					
Superintendent					
School Board					
Other Residents					

9. Has this component been subject to formative evaluation (Level One) by the school district?
 _____ No
 _____ Yes. Please describe. _______________________________

10. Is this component part of an attempt to fulfill new objectives or functions in the school district?
 _____ No
 _____ Yes (Continue with *a* to *d*)
 (a) What are the new objective or functions? (Describe) _______________
 ____
 (b) How much emphasis (priority) is being placed on this (set of) objective(s), compared to other objectives already being achieved in this area before the component was introduced?
 (Circle one)
 (1) Supplement(s) other objectives in only a minor way 1
 (2) Is (are) now of equal importance to other objectives 2
 (3) Is (are) now of primary importance, the other objectives having far less priority 3
 (4) Completely supersede(s) the other objectives 4
 (c) How different is this (set of) objective(s), from previous ones? (Circle one)
 (1) Very different 3
 (2) Somewhat different 2
 (3) Not very different 1
 (d) To what extent are these new objectives actually being fulfilled?
 (1) Fully 4
 (2) Largely 3
 (3) Partially 2
 (4) Not at all 1

11. Does this component involve the use of one or more new procedures/techniques/activities in curriculum; staff division; community participants; use of time and space facilities; organization, administration, and governance; and evaluation? (Indicate Yes or No in each column in the chart below)
 (a) Please consider *separately* each column for which you wrote "yes" and provide the following information within those columns using code numbers where appropriate. (Note: when the term "techniques" appears, it implies "procedures" and "activities" as well.)
 (b) How many new and distinct techniques can be identified? (Write the number and describe briefly)
 (c) Are these techniques *additions* to or *substitutions* for something that existed previously? (Write in the appropriate number. additions = 1; substitutions = 2)

(d) How much of the time is this (set of) technique(s) used in comparison/in addition to the former ones? (Write in the percent of the time)

(e) Which of the following apply to this (set of) technique(s)? (Write in the appropriate number)
- (1) Supplement(s) previously existed techniques in only a minor way 1
- (2) Is (are) now of equal importance to previously existing techniques 2
- (3) Is (are) now of primary importance; the previously existing techniques still used in a supplementary way 3
- (4) Completely supersede(s) the former techniques 4

(f) How different is this (set of) technique(s) compared to the previously existing ones? (Write in the appropriate number: very different = 3; somewhat different = 2; not very different = 1)

(g) How well does this technique work, that is, do what it is intended to do? (Write in the appropriate number: extremely well = 3; moderately well = 2; not well at all = 1)

	Curriculum	Staff division	Community participants	Use of time and space facilities	Organization administration and governance	Evaluation
(a)						
(b)						
(c)						
(d)						
(e)						
(f)						
(g)						

	Teaching experience	Professional reading	Staff modernity	Educational level	Father's educational level	Age	Percentage male	Level of tension	Frequency of disputes	Morale	Change orientation
Teaching experience	1.00										
Professional reading	.19	1.00									
Staff modernity	−.19	.20	1.00								
Educational level	.05	.15	.01	1.00							
Father's educational level	−.19	.25	.26	−.03	1.00						
Age	.60	.32	.01	.03	.08	1.00					
Percentage male	−.13	−.15	−.16.	12	−.27	.04	1.00				
Level of tension	−.30	−.01	.27	.11	.11	−.17	.17	1.00			
Frequency of disputes	−.27	−.23	.44	.07	−.04	−.16	.27	.52	1.00		
Morale	−.07	.19	.24	.28	.12	.23	.00	.58	.39	1.00	
Change orientation	−.13	.21	.13	−.04	.15	−.29	.07	−.06	−.27	−.23	1.00
Orientation to pupil autonomy	−.29	.19	.44	−.08	.33	−.43	−.06	.08	.02	.00	.66
Collegiality	.20	−.06	−.18	.09	−.22	−.07	−.20	−.49	−.51	−.38	.05
Perception of problem	−.23	−.04	−.11	.09	.15	−.08	.12	.43	.28	.24	−.17
Goal differentiation	.28	−.13	−.15	.11	−.20	.06	−.46	−.31	−.15	−.03	−.01
Goal discrepancy	−.35	−.28	.35	−.01	.16	−.23	−.35	.43	.61	.36	.02
Complexity	.27	.36	−.10	−.10	−.11	.28	−.04	.01	−.04	.11	.02
School size	.20	.08	.09	.43	−.13	.23	.14	.14	.24	.36	−.35
Formalization	−.01	−.17	.02	.18	−.05	.12	−.00	−.12	−.09	−.18	−.27
Individualized technology	−.07	.18	.10	−.12	−.05	−.32	−.48	−.16	−.06	−.05	.30
School level	−.15	−.14	.11	−.10	.02	−.01	.35	.05	.17	−.03	−.02
Classroom autonomy	−.28	−.32	−.11	−.20	.06	−.27	.24	−.37	−.15	−.47	.22
School-board authority	−.03	−.04	−.21	.41	−.07	.04	−.10	−.22	−.26	−.09	.10
Superintendent authority	−.02	.22	−.06	.38	−.08	−.19	−.18	.01	.03	.27	.16
Principal authority	−.23	−.27	.05	.12	−.26	−.01	−.00	.06	.17	.02	−.24
Teaching authority	.07	−.21	.16	−.10	.08	−.29	.08	−.19	−.04	−.42	.20

$r = .25$ **significant at the .05** level.
$r = .34$ **significant at the .01** level.

APPENDIX C

Correlation Matrix of Independent Variables (N = 45)

Orientation to pupil autonomy	Collegiality	Perception of problem	Goal differentiation	Goal discrepancy	Complexity	School size	Formalization	Individualized technology	School level	Classroom autonomy	School board authority	Superintendent authority	Principal authority	Teacher authority
1.00														
−.00	1.00													
−.20	−.62	1.00												
.01	.24	−.15	1.00											
.10	−.44	.30	.00	1.00										
−.01	−.16	−.04	.01	−.12	1.00									
−.19	−.21	.08	.02	.18	.53	1.00								
−.29	−.03	.16	.03	−.14	−.21	.03	1.00							
.59	.24	−.21	.14	−.18	−.08	−.35	−.34	1.00						
−.21	−.26	.25	−.22	.48	−.05	.12	−.01	−.35	1.00					
.07	.23	−.25	−.02	.02	−.07	−.16	−.03	−.12	.27	1.00				
−.11	.18	.13	.23	−.22	−.30	−.19	.18	.01	−.11	−.06	1.00			
.26	.33	−.37	.19	−.03	.24	.23	−.41	.25	−.25	−.06	.07	1.00		
−.10	−.09	.29	−.01	−.12	−.15	.01	.37	−.10	−.10	−.18	.32	−.03	1.00	
.35	.29	−.35	.18	−.03	.07	.04	−.03	.12	−.04	.33	−.18	.15	.06	1.00

Author Index

Subject Index